ONE SIGNAL
PUBLISHERS

ATRIA

BEFORE IT'S GONE

Stories from the Front Lines of
Climate Change in Small-Town America

JONATHAN VIGLIOTTI

ONE SIGNAL
PUBLISHERS

ATRIA

New York   London   Toronto   Sydney   New Delhi

ONE SIGNAL
PUBLISHERS

ATRIA

An Imprint of Simon & Schuster, LLC
1230 Avenue of the Americas
New York, NY 10020

First One Signal Publishers/Atria Books hardcover edition April 2024

ONE SIGNAL PUBLISHERS / ATRIA BOOKS and colophon are trademarks of Simon & Schuster, LLC.

Simon & Schuster: Celebrating 100 Years of Publishing in 2024

For information about special discounts for bulk purchases, please contact Simon & Schuster Special Sales at 1-866-506-1949 or business@simonandschuster.com.

The Simon & Schuster Speakers Bureau can bring authors to your live event. For more information or to book an event, contact the Simon & Schuster Speakers Bureau at 1-866-248-3049 or visit our website at www.simonspeakers.com.

Interior design by Joy O'Meara

Manufactured in the United States of America

1 3 5 7 9 10 8 6 4 2

Library of Congress Control Number: 2023950210

ISBN 978-1-6680-0817-1
ISBN 978-1-6680-0819-5 (ebook)

For my father,
who said one day I'd write a book

CONTENTS

PART ONE: **FIRE**

PART TWO: **WATER**

BEFORE IT'S GONE

Preface:
Ticktock

Mother Nature just wiped another American town off the map. It won't be the last, and (as you'll read here) it's far from the first.

I was completing edits on this book when I got the call at four a.m. "How soon can you get to the airport?" the voice on the other end asked.

This is the life I've chosen. And it's how—less than twenty-four hours after a wildfire ripped through postcard-perfect Lahaina on the Hawaiian island of Maui—my CBS News team and I found ourselves on a deep-sea fishing boat named *Nemesis*, carving a coast-hugging path through the swelling Pacific. Our choppy maritime trail was the only way to bypass police roadblocks that residents told me were less about protecting people from danger than preventing journalists from documenting a crippled emergency response and growing humanitarian crisis inside. We were on a reconnaissance mission. "The world needs to see whatever's going on," *Nemesis*'s pilot, Captain Tim, said with a toothless grin as he took a puff from a clove cigarette, his exhales sweetening the salty ocean spray. The thick specter of black smoke we were heading toward would soon reveal an environmental holocaust fueled by both human-caused climate change and the local government's failure to protect its people from our planet's radicalized elements.

As we got closer to ground zero, my producer Christian Duran, fixer Shanti Berg, and I went over the plan. The order was to prioritize. We'd be in triage mode once we touched down, with only thirty minutes to film, interview the affected, and gather other reporting before Captain Tim would be forced by dimming light and worsening wind to take us back to Maalaea Harbor, where we had originally set sail. It was a small window of time, complicated by all communications in the region being knocked offline by the fire. If we missed our rendezvous, we'd be trapped in this godforsaken crematorium, unable to report back to the CBS News bureau—and the public—what we were seeing. Unable to get this community real help. "Keep an eye on your watch. You don't want to be stuck out there at night," Captain Tim warned as we approached Lahaina's cindered shoreline a sea-sickening hour and a half later. "There's no power and we're hearing reports of looters."

It was just before sunset when we wobbled onto a makeshift dock on the outskirts of town. Through a strip of lush vegetation and vapor that curtained the coast, we discovered a miles-long field of smoldering ash and twisted metal. Flames radiated from inside the foundations of what were once homes, like thousands of small campfires in the final moments before bedtime. The stories we would hear in this flickering and smoky twilight will forever haunt me, along with the smell . . . an acrid cloud of burning metal, asbestos, wood, plastic, vegetation, skin, hair, and bone. "This is just the beginning," said a man who pulled up on a motorcycle. "I just came back from the other side of town. It's all gone," he said before taking off. Without a car, we focused on what we could walk and run to, and from what we saw, there was little left to pillage. Home after home, business after business, hotels and entire apartment buildings had been leveled. The man on the motorcycle wasn't exaggerating.

On foot and by drone, we recorded destruction and human desperation, but not a single first responder. No firefighters, no police, no strobing lights and blaring sirens signaling help was on the way.

"How could journalists arrive first?" I wondered out loud. Where was the relief? The few survivors we did meet phantomed this American Pompeii. Lines carved by tears in their ashen faces let me know they were very much alive. "It just came out of nowhere, like a ball of flames, and people were screaming, crying, and begging," said a man I ran into on the corner of Front Street and Mala Wharf Road. There was still so much more to document and so many questions that demanded answers, but our time was up. Hungers to learn more would have to wait until sunrise or we'd risk sleeping in Hell on Earth.

My team and I made several clandestine trips back to Lahaina in those early anarchic days where laws—along with the lives that lived by them—were reduced to charcoal and rewritten by those who survived. The following morning, unrestrained by darkness and deadline, Christian and I walked the entire two-mile oceanside stretch of Front Street. While there were still no cops, we did spot a small search and rescue team, dressed in head-to-toe white protective coveralls, combing through the ruins for the missing. To this point the death toll was only in the single digits, but thousands were unaccounted for and an ominous white van trailed close behind in our group's wake. We would later learn that before this crew began digging through the wreckage for fragments of bone and any identifying jewelry that may have survived the flames, they first collected entire bodies found on the streets and slumped over Lahaina's seawall. People had tried to outrun the flames but couldn't. Many passed out from smoke inhalation before the fire reached their heels. Some were literally only a few feet away from deliverance.

The extent of loss was hard to process as my mind tried to fill in the blank spaces with memories from before, when colorful shops and restaurants lined either side of the historic road. In 2019, while on the island covering a different story, I dipped into Fleetwood's for a cold IPA, bought a Hawaiian shirt at the Volcom store, and picked up some small gifts for my husband and friends at the Honolulu Cookie Company. Lahaina, population 13,000, was a quaint kaleidoscope of Polynesian culture that somehow charmed millions of tourists every year while holding on to its deeply rooted identity. Before Lahaina was the United States's paradise, it was the capital of the Hawaiian kingdom, home to King Kamehameha, who unified the volcanic archipelago under one name after defeating the other islands' chiefs in a fifteen-year war that ended in 1795.

On the afternoon of August 8, 2023, centuries of history were lost. Eighty percent of the town—more than 2,000 buildings—was destroyed. Christian and I were now walking through a no-man's-land, stopping to catch our breath, chug water, and rinse our mouths. The microscopic particles of Lahaina's incinerated DNA coated my tongue with a gritty film that had a burnt metallic taste and made it hard to breathe. We walked for hours. Every corner of town was frozen in time by fire and revealed a tragic story. Fleetwood's brick facade still stood, but its interior was gutted. The Honolulu Cookie Company and Volcom, Maui Jim, The Dirty Monkey, Pirate Jack's Tattoo, Cool Cat Cafe, Down the Hatch, Paia Fish Market—the list goes on—were erased down to their foundations.

Most startling though were the streets, seared by a gridlock of hundreds of blackened cars. In a desperate attempt to flee the flames, people quickly jammed the few arteries that led out of town to safety. But for most, it was too late. Christian and I wove through the skeletal remains of bumper-to-bumper traffic. We met one married couple inspecting what was left of a Nissan Pathfinder. The wife held its key fob in her hand. "We got stuck and had to jump into the ocean," she told us, looking at the clear blue Pacific, which lapped up along the seawall that marked the edge of Front Street. Her husband politely said goodbye as they continued tracing the steps that determined their fate that afternoon. As Christian and I continued, we passed a dog—a black Lab or a pitbull, it was hard to tell—lying dead in the street. Through the shattered window of a nearby car I could see what looked like a piece of jawbone on the metal springs of the burnt-out driver's seat. Its porous texture and organic shape stood out among what remained of the rigid manufactured car parts. "No," I said out loud, literally shaking the idea out of my head.

Down the road, across the street from where Lahaina's public library once stood—a humble single-story building that housed a maze of packed oak bookshelves—we met Carmelo Moran Garcia. He had returned to look for his friends. "Their apartment was somewhere over there," he said, pointing to an entire block of rubble. "I can't tell because the fire erased the streets. I haven't heard from them in days and I know they were home at the time."

As Christian and I reviewed what we filmed in the parking lot of a Safeway, I struggled to understand how something like this could happen. Since the turn of the century, as more and more wildfires erupted on the island, teams of firefighters had been specifically trained to snuff them out. And like all of the state's islands, Maui had a network of sirens designed for disasters like tsunamis, hurricanes, volcanic eruptions, and, yes, wildfires. If the sirens sounded, locals knew to immediately turn on the TV or radio for updates . . . and if all else failed, head into the streets to look for a possible threat. And yet by all accounts, this one caught everyone off guard. The sirens never sounded. People never knew there was a raging fire making its way to their homes, much less were given a head start to escape. "There was no warning" would become a common refrain from both survivors and local leaders along with "the fire just exploded," as if to imply it all happened so fast and there was nothing anyone could do.

But in the weeks that followed, it became clear there was plenty that could have been done to prevent this tragedy. We soon learned Maui county officials had hours and even days to issue warnings and evacuations to the public, but didn't. These same officials also never heeded warnings from scientists in the years leading up to what would quickly become the deadliest wildfire in modern American history. Hawaii governor Josh Green described the fire that devoured Lahaina as a "bomb" going off, suggesting there was no time to take action, but in truth there was a years-long countdown clock that our team helped trace back to the moment it first started its terminal ticking.

The year was 2014, and the state-funded Western Maui Community Wildfire Protection Plan warned climate change would bring less rain, more drought, and an increased fire threat to Hawaii. The one-hundred-page report recommended managing invasive grasses, reducing dry vegetation (known as "fuel loads"), creating defensible space, working with private landowners to build fire breaks, and finding ways to conserve water. Officials from the Maui Department of Fire and Public Safety, the Maui County Civil Defense Agency, and the State Department of Land and Natural Resources signed the plan during a ceremony on June 12, 2014. That wet ink was a commitment to implement the suggested changes that the federal government would help foot the bill for. But in the years that followed, county officials wouldn't provide evidence showing any such adaptive measures were taken. The same 2014 plan was brought up again after a fire ripped through Maui's upcountry in 2018, but in a cycle that will become all too familiar in this book, it was brushed aside. Meanwhile, the detonation clock continued its ticktock and residents in Lahaina were unaware they were living on borrowed time.

In Hollywood blockbusters, the Tom Cruises of the world start to sweat when the countdown to detonation dips below sixty seconds. In Maui, this metaphorical minute began when the National Weather Service issued a "high fire danger" advisory on August 4, 2023, four days before the Lahaina Fire. Meteorologists were specifically concerned about strong winds from a hurricane named Dora that would pass Maui 700 miles to the south. The advisories got more dire as the clock counted down, and yet Maui County did nothing to prevent catastrophe. Tom Cruise might as well have been on lunch break. On August 7, the day before disaster struck, the National Weather Service issued its most dire high wind and fire weather alert: "Red Flag: High fire Danger with rapid fire spread. Stay safe and be cautious." Ticktock. Ticktock.

This would have been an "all hands on deck" moment for the county, but as my team would uncover, the head of the Maui Emergency Management Agency, or MEMA—a man named Herman Andaya—was off the island in the hours leading up to the fire, attending a conference in Oahu, ironically dedicated to emergency preparation. Perhaps unsurprisingly, many of Andaya's emergency management counterparts from the state's other islands pulled out of the event to prepare their communities for the passing storm. And for good reason. There were still steps that could have been taken to prevent disaster. Ticktock. Ticktock.

On the morning of August 8, 2023, ten hours before Lahaina would go up in flames, the school district, heeding the National Weather Service's warnings, canceled classes because of dangerous winds. If only the county had followed this lesson and taken its own precautions. Gusts up to 80 miles an hour had already knocked over trucks, blinded streets with clouds of dust, ripped awnings off buildings, and brought down trees and power lines. At six a.m., one such fallen line sparked a small fire on the outskirts of town that crews declared 100 percent contained hours later. One firefighter told me how lucky they were to tamper the flames before they grew out of control (in a major tactical failure—and for reasons that are unclear and against agency protocol—not a single engine was left to monitor the burn site for signs of life). Ticktock. Ticktock.

These were now the final ticktocks for county officials to take action; for Andaya to ring Maui County mayor Richard Bissen and demand the power grid be deactivated. This was the time to start preemptively evacuating residents and staging firefighters around the town. This was the time to activate those sirens so everyone was aware of the threat and could seek information and decide what was best for their families. This was the time to do a lot of things that were never done. Instead, many children sat at home alone, or with grandparents, as their parents worked. On Front Street, tourists shopped along the historic district, marveling at the high winds but otherwise unaware of the "100 percent contained" fire that suddenly began smoking again.

Unlike bombs—at least those in the movies—there is no magical red wire one can cut in the last few seconds to deactivate a storm. The strengthening wind from Hurricane Dora that swept Lahaina had found a hot ember hidden in the dry grass and blew it into a blaze just before three p.m. The smoke first appeared as a thin gray ribbon. Had fire crews been monitoring the burn scar, they could have quickly put it out. Heck, if fire crews were in a five-mile radius of the blaze, they could have arrived in time to get the upper hand, but in yet another tactical failure, most of the Maui Fire Department had been dispatched to a fire on the other side of the island, despite the clear threat to Lahaina. While at this point county officials had run out of most options to contain the growing fire and protect structures, there was still time to save lives. The yearslong countdown had hours left until *BOOM*. In fact, it would take two hours for the fire to burn in brush before "exploding" into a mushroom cloud–shaped monster that devoured land and life at a mile a minute. Two whole hours—time that could have been used to issue evacuations and clear people from the line of fire. Yet no such order was issued. As the fire engulfed Lahaina and its people, Mayor Bissen went live on local news delivering what he called "good news." The road to Lahaina was open, he said, seemingly unaware most of the town had fallen to flame thirty minutes earlier.

In the weeks that followed, we'd also learn what happened to those sirens. Initially officials, including the governor, mayor, and Maui County fire chief, said the alerts would have saved lives but were damaged by the wind. In the end it turned out the sirens not only remained in working order throughout, but Herman Andaya, the man who held the key to this fire alarm, intentionally did not activate it.

"You made the decision not to sound the sirens which could have saved many lives. Now there are questions about your lack of experience, that you only took online courses for disaster relief and had no in-the-field experience [according to LinkedIn, Andaya's previous jobs only included public relations]. Do you regret your decision, and have you considered handing over the reins to someone with more experience?" I asked at a televised press conference that was beamed live into homes around the globe. There was an audible gasp from those in the room. "To say I'm not qualified is incorrect," Andaya said before I followed up, asking again if he regretted his decision to not activate the sirens.

"I do not," he said defiantly. His response shocked me.

"So many people have said [others] could have been saved if they had time to escape. Had the siren gone off they would have known there was a crisis emerging . . ." I said as Mayor Bissen stepped in to defend Andaya. "Do you want him to give you the answer?" he asked.

"I do, but I want to get it out there. There's a lot of people who want answers," I said firmly.

Andaya went on to give conflicting reasons for his decision, including falsely asserting the sirens were only activated for tsunamis, even though the state's own website said they're also used for wildfires. He then said he worried people would run into the fire instead of away from it. "Does he really think we're that stupid?" survivor Shannon I'i later told me. But the icing on the cake, or the "crap on the ground" as resident Christine Borges put it, came moments later when Andaya claimed most people wouldn't have heard the sirens even if they had been sounded, even though they're so loud they rattle buildings for more than half a mile.

In less than twenty-four hours Andaya would resign due to "health reasons." Officials, including the head of the state's emergency response office, Herman Andaya's boss, went on to describe a local government response that was asleep at the wheel. The county hadn't even alerted the state about the extent of the crisis until the next day, when it was too late. "I thought everyone had gotten out safely," Major General Kenneth Hara, head of the Hawaiian National Guard, said. Not only had the county failed to pull the fire alarm, but they hadn't called on the state or federal government for help. Soon, the true toll of years, days, and those final hours of negligence would take shape.

"We're finding bodies anywhere there is a bed," Todd Magliocca told me in the field. He was helping lead the federal search and recovery effort. I was stunned. "People were asleep?" I asked. "People were asleep. People were in various stages of fleeing their location."

Those "various stages of fleeing" included entire families found in homes holding on to each other. Some victims were discovered inside dumpsters, hoping the metal walls would shield them from the flames. Many others were found in that gridlock of cars. The chief of police, John Pelletier, a brash man who started every press conference asking for "patience, prayers, and perseverance," described the horrific conditions in which his officers were finding victims. "When we pick up the remains and they fall apart, that's what we're dealing with . . . we've got an area that we have to contain that is at least five square miles, and it's full of our loved ones." In the end, he said, there would be a list of "confirmed dead and those presumed dead." The fire burned so hot, he cautioned, it would be impossible to locate every victim.

Then there were the victims that survived but were left to fend for themselves, like Kawena Kahula, a manager of a hotel in Kaanapali just outside Lahaina. For three days it was her job to keep hundreds of guests calm as they waited for help. Cell service, the internet, and landlines were all down. The power was out and there was no running water, but at least they had a sturdy roof over their heads. It was safer, they believed, to shelter in place than to head out into the unknown. In only the first twenty-four hours, hotel staff were forced to ration food, giving what little was left to the children. "There's an airport right up the road. Why isn't that being used? There's an oceanfront twenty feet from our lobby. Why are we not using that," she began to wonder as days went on with no sign of aid. When Kawena eventually decided to risk it all and drive down the one road out of town, asking God to protect her with every click of her car's odometer, she was surprised to find the path clear of trees and power lines. By the time she pulled up to the police roadblock thirteen miles later—the same roadblock my team and I bypassed by boat—she hadn't seen a single first responder and was frustrated to discover a line of volunteers with supplies not being allowed through. She felt more like an escaped evil spirit from Perdition than a survivor from a town brought to its knees by human-caused climate change and government inaction.

I hammer out this timeline for a reason: because behind the many mistakes that were made before, during, and after the fire that destroyed Lahaina lie solutions for how to protect other main streets across America from similar destruction. Following the inferno, Hawaii officials evoked "climate change" more than I've ever heard before while on the front lines covering extreme weather. But "climate change," in this context, sounded more like an excuse than an existential threat and a challenge we must all rise to. The very climate science Hawaii officials had ignored for nearly a decade was now revered in what appeared to be a strategy to place full blame on Mother Nature, while shifting focus away from human error.

Let's get something clear right now: Since our planet's birth, its climate has changed. But in the past this change was natural and slow—over thousands or millions of years. So slow, life on Earth largely found ways to adapt. But today, thanks to human-emitted greenhouse gasses like carbon dioxide and methane, those thousands of years of change are unfolding over mere decades, faster than Earth's inhabitants can keep up. As we'll explore in the pages to come, it will take many years to cut our emissions to the point where we can roll back the deadly consequences of the climate we've changed. But we're not helpless in the meantime. We can strengthen our communities and be ready for future storms. But in order to do so we must first listen to the warning signs . . . and take action. Which brings me back to the aftermath in Lahaina.

While Governor Green called the fire the "deadliest natural disaster" in state history, humankind also played a role. And not just by spewing gasses into the atmosphere that, like the greenhouse they're named after, heated up the Air, dried-out the Earth, and boiled our planet's Water, which in return fueled the kinds of wind events that turned a spark from a downed power line into a Firestorm. Sure, that's exactly what happened in Lahaina. But then there's also man's impact on the land. More development around the island meant less water to go around, which dried up lakes and marshes and turned a vibrant ecosystem managed by Native Hawaiians for centuries into a deadly colonialized tinderbox that placed tourism first. Yes, human-caused climate change fueled this inferno, but *habitat* change—and people's failure to recognize our radicalized elements as a threat—was the second (always less spoken about) accelerant.

Two weeks after Lahaina burned, President Joe Biden toured the disaster zone from the sky in a helicopter and on foot. It was a carefully choreographed visit that ended at a podium staged in front of the town's 150-year-old banyan tree. The "Great One," as some locals call it, is a cherished landmark in the community. Its lush canopy of waxy emerald leaves shaded an entire block and generations of families who gathered under it to escape the heat and marvel at the birds that called its snaking roots and limbs home. While flames torched the Great One, it was found unbowed. Many of those emerald leaves even managed to hold on, albeit in a rusted state, and they crackled in the breeze as the president declared the banyan's resilience a symbol of the town's resolve to rebuild. He said Lahaina would rise again and promised native Hawaiians a seat at the table during the process. Their ancestral land would not be snatched up by predatory developers looking to build all-inclusive hotels, he would later tell some residents during a visit to a donation center. It was exactly what residents wanted to hear, but not what they needed to hear. Yes, Lahaina will rebuild again, just like every other American town lost before it. But unless changes are made, another countdown clock will turn on, and it's anyone's guess when time will run out. Because before every "unprecedented" explosion in the cities and towns of a nation now under siege from an environment it spent too long taking for granted, there is a history of missed opportunities.

Prologue:
A World Away Hits Close to Home

My boss Andy Clarke had, over the course of four years working together, honed a familiar sequence of events I knew to follow as soon as word came in that I was needed to deploy to some far-off stretch of the world as a foreign correspondent for CBS News.

The CliffsNotes version goes something like this: grab passport, book flight(s), confirm "go-bag" is properly packed, double-check visas, repack poorly packed "go-bag," withdraw plenty of cash from ATM, rush to London Heathrow airport, calm nerves with a stiff drink on the plane, rendezvous with team at destination, and hit the ground running, while occasionally pausing for a guilty cigarette.

It was a rapid-fire, adrenaline-fueled routine that launched me into orbits I used to dream about as a kid flipping through my parents' old *National Geographics* in the basement of our home in Mount Kisco, New York . . . places like Essaouira, Colombo, the Gaza Strip, Nairobi, Sydney, Reykjavik, Ittoqqortoormiit, Ikaria, Istanbul, Dublin, Rome, Fraser Island, Pamplona, Paris, Edinburgh, Brussels, Copenhagen, the West Bank, Snowdonia, Suva, Tel Aviv, Beirut, Dubai, Auckland, Stockholm, Jerusalem, Tres Fronteras, Lofoten, Marrakesh, Nice, Crete, Gibraltar, Munich, Amsterdam, and Tonga, to name a few. This assignment, though, was different.

Andy's email that morning had been terse, a knowing choice of economy because the subject line said it all: "Come to my office. Let's talk Syria." I had begged Andy to send me there since the first day I walked into the newsroom as an in-over-my-head freshly shaved "kid from the States" and had been, in so many words, told to *piss off* each time. It became a running joke in the breakroom, where I'd often circle the kettle waiting for Andy to make his morning cup of tea, light with no sugar. "Send me to Syria," I'd say. "You think that's a good idea?" he'd respond as he walked away, not waiting for an answer. And fair enough. My ambition has always outpaced whatever skill it is I think I possess. But in television news, ambition is sometimes enough . . . Most likely I was tapped because none of my more respected and polished correspondent colleagues could get a visa in time. Fuck if I cared, though. I was finally headed to the front lines to document a civil war between men, and a pristine example of how the environment we all helped radicalize can lead to violent bloodshed.

The year was 2019 and the proof linking climate to conflict was piling up. "The evidence is clear that climate change does contribute to increased conflict, but along indirect pathways," the United Nations posted on their website. "There are a variety of context factors—in particular, socioeconomic conditions, governance, and political factors—that interact and play a key role in translating climate change into conflict risks." This deadly cauldron of ingredients had already boiled over in places like Sudan, where back-to-back droughts that first gripped the African nation in the 1980s triggered a domino run that began with famine and ended with political unrest and war. Today, five million people—or 40 percent of the population—still face extreme hunger and conflict. The dominos that led to Syria's demise toppled later, but along similar "pathways."

That's how I found myself in the village of al-Suqaylabiyah, walking through rows of the dusty olive trees that framed stone houses hammered into anthills by a wartime mix of mortars and shells. I arrived toward the end of Syria's climate chain reaction. It was early spring, and Syria's then-eight-year conflict had transformed every corner of the country. The muddy and fractured farmland I hiked through with a Kevlar vest and helmet had become the final flashpoint in the war as government forces closed in on the last rebel stronghold in Idlib.

British producer Barny Smith, Italian photographer Federico Pucci, and I traveled here with the National Defense Forces (NDF), a pro-Syrian government militia whose barracks were a collection of white canvas tents hidden in the orchards. As the sun set I could make out the silhouette of one fighter resting inside, sipping something hot from a mug. The NDF's targets, an estimated 70,000 rebels, remained in hiding among three million civilians, according to the government, and while a ceasefire negotiated by Russia and Turkey had minimized bloodshed, tensions were still rising. Innocent people—men, women, and children—became pawns, caught in the crossfire of bombardments. Meanwhile, in the eastern part of the country, U.S.-backed Kurdish rebel forces closed in on the ISIS caliphate, the terrorist group that filled the power vacuum created when the Syrian War first began in 2011. As the battle for control of this land waged on, the slices of the pie that had already been ripped back by deadly force struggled to rebuild.

The signs of death could be seen everywhere as my team and I drove through cities like Homs, Aleppo, and Palmyra. It had been two years since the Syrian government and its allies "liberated" Palmyra, and only thirty-five families had returned, among them Feda Mahmoud and her son. "I've been living here for thirty years, so I don't want to leave," she told me with the help of an interpreter as she grilled chicken on the street in front of her small cafe. A hijab covered Feda's round face. Her large, thick hands dwarfed the kebabs she served to the hungry. Those mouths mainly included the Syrian-allied Russian and Iranian forces still stationed there and, on this day, mine. As I walked through the city that afternoon, I looked through the blown-out doors and shattered windows of homes and street-level businesses that I imagined resembled Clinton Street on New York City's Lower East Side, where I once lived while working as a reporter for WNBC.

War erased Palmyra down to its cinder-block frames, whose exteriors were pockmarked by thousands of bullet holes. An entire city of 200,000 people vanished without a trace, and without a reliable source of water or electricity, it would take a special breed of pioneer to rebuild. "I hope that everyone one day will come back because there's something we say: 'Even in heaven, if there's no people, you can't sleep,'" Feda told me earlier.

President Bashar al-Assad described the devastation at the time as "painful for us as Syrians to see," and justified his army's bombing campaign to eradicate rebels from cities like Palmyra. "This is the price sometimes, but at the end the people are liberated from the terrorists." Critics called this "reclamation" a genocide.

What can't be refuted is the lingering toll of this conflict. According to the Syrian Observatory for Human Rights, nearly half of Syria's prewar population have been displaced and more than 600,000 people have been killed. "It's a cancer, what's happened here. And to think all of this destruction began with desperation in our farms," said Zaher Sabouni, a soap maker who used to source his olive oil from Syria's rural "breadbasket," where the first domino fell. While bombs brought Syria to her knees, the region's rapidly shifting climate and mismanaged land—overlooked for decades—collided, causing unimaginable pain to the people who remained.

To understand the threat a radical environment poses to nations, and Mother Nature's role in Syria's fall, you'll have to rewind through all the destruction to a time before the war began. Before the bombs and fighter jets and shattered puzzles of concrete. The year was 2006 and Syria's cities were bustling, blissfully unaware of the emerging crisis in the country's agricultural veins. The farming community of Daraa was once a cultivated land of wheat and cotton, but five years before civil war would erupt, this fertile Eden began to unravel for a number of related reasons that had gone ignored.

Okay, let me pull over to the side of this metaphorical road and explain something. Us humans are essentially tiny microorganisms— bacteria, fungi, and the like—living in Earth's planetary body, also known as the biosphere. In this biosphere, our climate (weather) and our habitat (including trees, plants, and rivers) act as vital organs regulating four key elements: Air, Water, Fire, and Earth. Like blood, these elements circulate nutrients through the entire body. But decades of human activity have fueled changes in our climate and habitat that have radicalized the behavior of our elements, and in return triggered environmental disasters the likes of which have never been recorded in modern, developed history. One by one, rural, suburban, and urban ecosystems—the biosphere's fingers, ears, and toes in this allegory—have gone gangrenous. I tell you all this because many people consider "climate change" the only threat we face as a society, while overlooking the other contributing factors in our environmental crisis. While our climate does play an outsized role, it's not alone, and by focusing only on *it* we fail to see the many other options that exist to turn things around.

In the pages ahead, I use "climate change" when specifically talking about our warming atmosphere; "habitat change" when talking about damage to our land, waterways, and wildlife; and "ecosystem collapse" to describe the changing climate and habitat's combined role in knocking main streets around the world offline. Which brings me back to Syria.

Centuries of humans all over the world spewing heat-trapping greenhouse gasses into the atmosphere, along with poor irrigation and land management practices and one of the worst long-term droughts in modern history, had scabbed this once verdant land in brittle layers of condensed dust. Farmers in Daraa, the agricultural center of the country, warned the government of a looming collapse for years. Seeds wouldn't sprout and families were barely hanging on. If aid wasn't provided, they said, people would abandon the fields and flood already packed urban ecosystems, putting too much stress on limited resources and uprooting life in the nation's agricultural land. It was a self-fulfilling prophecy. By 2007, 60 percent of the nation was hit by extreme drought and 75 percent of farmers suffered total crop failure and livestock mortality, according to a report published by the *Proceedings of the National Academy of Sciences*. The crisis only worsened as time marched on, hollowing out community after community. An estimated 1.5 million people abandoned rural areas for urban ones. "Down with the regime. Your turn next, Doctor" then-fourteen-year-old

Mouawiya Syasneh graffitied on a wall in Daraa's downtown, taking direct aim at President Assad, a trained ophthalmologist.

What Mouawiya would later call a message of "desperation" triggered a brutal backlash from government forces, and the protests that erupted in response ignited one of the deadliest wars in modern times. Daraa's damaged climate and habitat had fueled environmental disaster that, without adequate government intervention, eventually led to ecosystem collapse. The loss of local identity along with poverty, pain, and "desperation" were a radicalizing force of unimaginable power. Syria's "breadbasket" became "the cradle of the revolution."

"Weather extremes, such as meteorological droughts, have been found to sometimes coincide with armed conflict outbreaks . . . Agricultural production shocks, often with subsequent migration, constitute the most commonly suggested mechanism linking climate to violent conflict," an international group of researchers concluded in the 2022 study "Societal Drought Vulnerability and the Syrian Climate-Conflict," published by the science journal *Nature*.

While Syria's "climate conflict" presents an extreme example, it also must serve as a lesson for what other countries in other parts of the world could face if action isn't taken now. I've covered ecosystem collapse for much of my career as a journalist, but weeks spent tracing the origins of Syria's deadly war was the kind of assignment that rewires a journalist's inner circuit board. I landed back in London with a stubbled face and fresh eyes.

———

A few months later, I moved to the network's West Coast bureau in Los Angeles and began searching for Daraas closer to home (this was a move planned months earlier, but we'll get to that soon). In the time between then and now I have covered historic hurricanes, thermometer-shattering heat waves, record-breaking droughts, megawildfires, back-to-back "hundred-year floods," unprecedented blizzards, and never-before-seen mudslides. According to the National

Oceanic and Atmospheric Administration (NOAA), between 2020 and 2023 the United States was hammered by more than *eighty* billion-dollar environmental disasters that destabilized highly developed communities that, as it turns out, were poorly adapted to survive today's storms.

I was documenting the early signs of ecosystem collapse destroying American main streets in ways so many had struggled to imagine happening here. And I understand why. The unrivaled industry, infrastructure, and engineering that have helped stress the United States's ecosystems have also, paradoxically, shielded most of us from the elements we radicalized. But now nature is outmaneuvering man's ability to hold her back, and our modern tools can no longer spackle over the growing cracks. "As families risk their lives in search of safety and security, mass migration leaves them vulnerable to exploitation and radicalization, all of which undermine stability," said then Secretary of Defense Lloyd Austin during the 2022 Leaders Summit on Climate. He was describing seismic change in Africa and Central America, but the United States sits on similar fault lines. Along the main streets of the ecosystems that form America, more people than ever before are feeling the impacts in real time. Not a century from now, not in a decade, but right now.

The real estate analytics firm CoreLogic found, in 2021, 14.5 million American homes were affected by environmental disasters. That figure works out to one in ten homes in the country. More than a million people were forced to either permanently or temporarily relocate because of extreme weather and habitat destruction. Research conducted by Forbes Home also found nearly a third of Americans surveyed cited "climate change" as a reason for moving in 2022. This eco-migration is straining both host communities and the communities left behind.

The social and cultural conflict following Hurricane Katrina was the example at home we Americans seemed to forget. On August 29, 2005, a hurricane that exploded into a Category 5 made landfall in New Orleans, exposing both humanity's role in birthing the twenty-first century's new breed of megastorms, and our inability to hold Nature's radicalized elements back. The surging water of the gulf toppled levees and ripped through one of America's capitals of black wealth and culture. Eighty percent of the city was flooded with up to 20 feet of water, and an estimated 1,833 people were killed. But Katrina's devastating ripple effect would soon make bigger waves around the country. According to The Data Center, a New Orleans–based research organization, more than one million people were displaced by the storm—many of them people of color—and forced to find refuge in other communities both near and far. But not every climate refugee was welcomed with open arms, and the data from this post-storm migration is devastating.

Research from Georgetown University showed people in villages, towns, and cities that received climate refugees from the Gulf States impacted by Katrina's landfall became "less supportive of spending to help the poor and African Americans." Some cities actually pushed for more spending on crime while opposing funding for public benefits. It's similar to what happened in Europe, where many Syrian refugees fled. This race and class clash is unfolding more publicly today along the Mexico border, where millions of migrants arrive each year, many of them fleeing ecosystems ravaged by everything from hurricanes and severe droughts to failed crops. Back in the United States, abandoned main streets also struggle to survive without taxpayers. Residents that do remain in these stressed ecosystems show a growing distrust in the local, state, and federal government's ability to protect them, according to a Rice University paper titled "Under Pressure: Social Capital and Trust in Government After Natural Disasters," first published in the journal *Social Currents*. "As climate change continues to escalate the threat from natural disasters, government officials and disaster managers must pay attention to these negative impressions and focus on ways to build trust so communities can recover more quickly," the study suggests.

President Donald Trump first tapped this toxic stew of distrust, racism, and classism when running for office in 2015, before eventually weaponizing it on January 6, 2021, in his own desperate attempt to hold on to power. While the insurrection included extremist groups like the Proud Boys and the Oath Keepers, research by the University of Chicago found a majority who stormed the Capitol that day were acting alone but were radicalized by the same "fear that rights of Hispanic people and Black people are outpacing the rights of white people." By analyzing data provided by the Department of Justice and cross-referencing charging documents with local news reports and background checks, I found many of the more than 1,000 rioters arrested came from or lived close to ecosystems adversely impacted by migration and extreme weather events, including wildfires, floods, and crop failure caused by droughts. In fact, enough of the perpetrators had agricultural links that farm groups like the National Milk Producers Federation (NMPF), the American Farm Bureau Federation, the National Young Farmers Coalition, and the National Farmers Union (NFU) all separately condemned the attack. "Yesterday's insurrection put thousands of lives in danger in a brazen mob attempt to disrupt the peaceful transfer of power," NMPF president Jim Mulhern issued in a statement. "These acts of intimidation and terror have no place in this country and they cannot be condoned or brushed aside," wrote NFU president Rob Larew. And we should not brush aside what I believe to be one root cause. Human activity has radicalized our elements and our ecosystems—and in some cases the people that live in them—in ways we don't even recognize.

I've been to all fifty states—many of them many times—and have walked through prewar Daraas by different names like Dexter, Dawson Springs, Palmer, Paradise, Red Lodge, Leesville, Breezy Point, Eastham, Sugarloaf Key, Fair Bluff, Mayfield . . . the list goes on. Today the geometric crisscross of America's main streets forms the cradle of a new revolution pinning humans against Mother Nature. And the main street of every ecosystem has a Mouawiya sounding the alarm. In the pages ahead, I'll introduce you to people like Becky James, Jim Kalbach, and Eva Dawn Burke—everyday Americans I've met while documenting climate change and its role in our extreme weather, each struggling to survive in a habitat threatening to erase their ways of life.

Their accounts are enlightening, heartbreaking, and inspiring. While they all come from different ecosystems, collectively they illuminate a mounting domestic crisis and form a blueprint for adapting, surviving, and even thriving in our changing Americana. But we have to listen, and then collectively take action. *Before it's gone.*

Author's Note

The following pages have been broken into the four elements that human activity has radicalized—Fire, Water, Air, and Earth. Each section tells the man-made story of how the corresponding element became radicalized, the impact that element is having on our weather patterns and habitat, the scientists that first sounded the alarm, the communities stressed when action wasn't taken, and the pioneers now paving a new path forward.

Not all elements are created equal, and Air is the alpha of this group, influencing the rest more than any other. While this is probably where I should begin, I don't. This book follows my journey through the main streets of American ecosystems, and Fire is where it all began for me. So that's where we'll kick things off. But as you read, remember: Most if not all environmental disasters are fueled by more than just one element. For example, landslides and floods can be and are triggered by Air, Earth, Water, and Fire—both alone and in deadly concert. And many of these disasters are intertwined and have a compounding effect on one another, like how extreme heat causes and amplifies droughts, which stoke wildfires that in return can cause and amplify flooding, landslides, and mudslides (I'll explain this later). I try my best to place each disaster under the primary elemental driver. At its core, I hope this book provides an intimate snapshot of different American ecosystems at a crossroads.

And on that note: I have always believed the lens of a camera—and in this case the written word—is a window into understanding distant worlds and a mirror for reflection and analysis, but after reporting for nearing twenty years, I've learned some people don't want to see what's happening right in front of their eyes. I've had cranky viewers find me on Instagram and X, formally called Twitter, and leave wonderful messages like "Hey dipshit. Explain exactly how did 'human caused climate change' cause the wildfires in California. Exactly where did you get your PhD . . . Dumbass University?"

Funny enough, that person has a point. Because as you're already starting to see, our changing climate alone isn't responsible for wildfires. Also, I'm not a scientist. I studied journalism at Fordham University in New York City and have used my platform since then to expose critical issues impacting the natural world with the help of leaders in their respective fields. Journalists may not know everything, but I'd like to think we're pretty damn good at tracking down the people who know a lot, and breaking down their sometimes overwhelming knowledge into digestible pieces. CBS News alone has easily invested more than a million dollars in training and sending me and my team into the field in pursuit of the truth. It's been a priceless education. For some though, I realize none of this merit matters. They'd rather get their information from their crazy uncle on Facebook. Good for them. Those people can still reach me on Instagram and X. But to rip a page from Greta Thunberg, I, too, can also be emailed at smalldickenergy@getalife.com.

PART ONE

FIRE

The Flame Tamer

The small reception area of the United States Forest Service's Missoula Fire Sciences Laboratory—the one where visitors pass through on their way to secured rooms where researchers experiment with fire—looks like a mid-century still-life of bureaucratic perfection. The walls are wrapped in whiskey-colored wood paneling. A three-seater cream-colored couch is flanked by end tables made of the same warm, grainy wood, and above them sit wood-framed glass boxes—the sort of displays you'd see in a natural history museum. Inside are gadgets labeled with names like "aspiration psychrometer," "visibility meter," and "double tripod heliograph." A painting of their inventor, a man friends called "Gis," hangs behind them. It's the first thing guests see as they walk into the room. The portrait's emerald-green backdrop—thinly brushed so you can see each oily stroke on the textured canvas—is the only pop of color in the otherwise varnished amber space, but it is Gis's eyes that grab everyone's attention. They are soft and kind and contradict the rest of his thin, sharp face. His deep frown lines, combed-back graying hair, caterpillar eyebrows, and matching thick mustache give him a stern look that is further accented by his brown suit jacket, beige button-down shirt, and gray tie. He appears more like a high school English teacher nearing retirement than what the plaque under his portrait reads:

Harry T. Gisborne
Pioneer Forest Service Fire Scientist

When Harry Gisborne joined the Forest Service back in 1918 he was best known as the son of a timber family from Montpelier, Vermont, and a graduate of the University of Michigan's Department of Forestry, where he studied the emerging threat of wildfires. As the Forest Service's first fire scientist, Harry needed to prevent a repeat of 1910.

Fire season started early that infamous year, sparking first in northeastern Montana's Blackfeet National Forest on April 29. By June, the woods glowed an eerie orange in a hundred different places from Montana to as far west as Washington State. Wildfires at the time were mostly of little concern. They were typically ignited by lightning in remote areas far from humans and were considered critical to forest ecology for clearing out dead vegetation to make room for new growth. There wasn't even a single person in the country specifically trained to fight them. But decades of building new mining and logging towns in wildfire country where people really didn't belong were nearing a tipping point. The rapidly developing West, built on top of a seasonal tinderbox, was about to explode.

By early August, as many as 3,000 fires lit up mountainsides like constellations in the night sky. While lightning had sparked some, most had been accidentally ignited by loggers and homesteaders. Others could be traced to the freshly laid steel web of railroads carrying coal-powered locomotives that were known to spew red-hot cinders in their wakes. Mankind had unwittingly reengineered the West's natural habitat, with disastrous results. The U.S. Forest Service, then only five years old, was the federal government's solution for maintaining the country's increasingly treaded-on land. The agency's prime objective was to protect clean sources of water and manage healthy stocks of timber for a growing country hungry for both—"for the greatest good, for the greatest number, for the longest time," as they put it. But there was no plan in place to deal with wildfire. Civilian firefighters, men who were only trained to fight structure fires in cities and towns, were sent to the front lines to help, along with drunk volunteers coaxed out of bars with the promise of a day's pay and a future night of drinking funded. To no surprise, they were all in over their heads.

At the request of the agency, begging really, President William Howard Taft deployed around 4,000 soldiers to assist, and in the end the military might appeared to pay off. Most of the flames were under control in a few weeks. On August 19, soldiers were told to pack up and head home—job well done. But primitive weather forecasts at the time couldn't predict what was to come just twenty-four hours later. Hurricane-force winds swept in around one p.m. on Sunday, August 20, first hitting eastern Washington where hotspots grew into swirling flames that looked like the crashing waves of an angry ocean. The 70-mile-an-hour gusts pushed this fiery current farther eastward, reigniting the brush in Idaho and Montana. The flames grew in height and intensity as they fed off each new acre of dry vegetation. Some firefighters described the blaze as 200 feet tall. In just forty-eight hours, the wind whipped the smoldering remnants of thousands of tiny fires into one giant monster that devoured three million acres. "By noon on the twenty-first, daylight was dark as far north as Saskatoon, Canada, as far south as Denver, and as far east as Watertown, New York. Smoke turned the sun an eerie copper color in Boston. Soot fell on the ice in Greenland," reported the U.S. Forest Service. On the third day, the rain came and put out the so-called Great Fire of 1910 as quickly as it began, but not before it had destroyed several towns and killed eighty-six people. Many of the victims were those untrained civilian firefighters and barflies. Incredibly, most of the devastation happened in just a six-hour period.

Harry's War

Harry was only seventeen years old when headlines like "Towns Disappear in Sea of Flames" were splashed across the front pages of newspapers at the country market down the street from his family's home. He would later tell colleagues his path to the Forest Service was partly inspired by the 1910 catastrophe. Harry was a conservationist at heart and knew wildfires were a natural part of the landscape and good for forest ecology, but he'd been hired here for a different task. Politically, the pressure was on to hold back the flames. It may come as no surprise that the powerful timber and mining unions, along with taxpayers in emerging woodland towns, didn't like to see fire in their backyards. America was also in a period of rapid development, a crucial component of which was the wood used to construct new homes and businesses. Scorched timber wouldn't cut it. To keep Washington, D.C., happy and to keep the federal funding flowing, the Forest Service would embark down an irreversible road, bending Mother Nature to man's will in what is known today, in rather bureaucratic terms, as the "10 a.m. Policy." Every fire must be put out by ten a.m. the next day.

Ten a.m. was Harry's deadline each day at work as he rallied to help protect those in danger. He would need to identify wildfire conditions, track weather capable of sparking and driving a fire's growth, and offer guidance on fire behavior and techniques to prevent and suppress it, which would eventually be used to train the nation's first wildfire battalions. Paradoxically, the better Harry and his counterparts got at their jobs of wrestling nature into submission, the worse citizens got at recognizing the inherent dangers of moving into these wild lands. A false sense of security became commonplace, and ever more Americans put themselves in harm's way.

The American West of 1918 was still uncharted territory. "We have nothing to help us except our own imagination and what little ingenuity we possess," Harry wrote from his new office in the Forest Experiment Station in Priest River, Idaho. According to U.S. Forest Service archives, he launched here and there to freshly sparked flames, meticulously documenting their behavior and the conditions surrounding each blaze. The worst fires, he soon discovered, shared three causes: heat, drought, and wind. He believed that when each of the three sides of this "fire triangle" passed a specific threshold of intensity, an inferno was only a spark away. Working with the Forest Service's Forest Products Laboratory, he created those now-antique gadgets kept in wood-framed glass boxes to measure atmospheric humidity and moisture content in vegetation at the scenes of active fires. He then used his findings to monitor for similar conditions in unburned land, as documented in his report "Measuring Fire Weather and Forest Inflammability." Harry was like a doctor performing a physical on America's forests. The higher the blood pressure, the greater the threat of a critical incident.

He also developed methods of tracking what kinds of storms could be expected to produce lightning and, more significantly, wind (wind acts as both a fan and source of fresh oxygen—and fires, like humans, need it to survive). Meteorology was virgin territory in the period, but Harry's thinking was straightforward: If you had enough people trained to identify weather conditions and the direction in which a storm system was moving, it would be possible to forecast the weather days in advance. By the 1930s, he had built a network of more than 6,000 weather watchers across the country—an analogue version of today's radar. Harry's research meant communities could be warned about elevated threats of fire, and firefighters, once trained, could be put in place to attack a blaze in the event one was sparked by man or nature. The year 1934 would provide a severe test of his findings.

Readings from an Idaho ranger station indicated perfect conditions for fire. Harry's web of observers identified a storm system in approach, with high winds almost certain to wick any potential ignition points. He raised the alarm and suggested a mobilization of civilian firefighters and soldiers to head off the storm's impacts, but his request was denied. The Pete King and McLendon Butte fires, as they were later called, burned for months, stanched only by winter snowfall. In the meantime, several towns were destroyed as the fires swept over 250,000 acres. So much for the "10 a.m. Policy." It was the last time the federal government questioned Harry's work.

In the years that followed, the "fire danger rating system"—which tested field conditions and then ranked the likelihood of a fire on a scale of "low" to "extreme"—became the official way to prepare communities for the possibility of a fire. Today you can see the "lab results" from Harry's system on a chart displayed and regularly updated outside most national parks and forests. It's the one with the cutout of Smokey Bear reminding visitors that "Only you can prevent forest fires."

Harry's field research, including what he observed in the Pete King and McLendon Butte fires, also revealed a pattern in fire behavior. He noted, for example, that roadways, rivers, and areas with little vegetation were effective in preventing fires from growing. Without a fuel source, fires had nowhere to burn but out. The key to suppressing fire was to create what he called "control lines," yards-wide trenches in the earth, around a growing blaze. It was a lot like corralling cattle, except Harry's cows were towering flames. The discovery helped pave the way for training the country's first official wildland firefighters. The civilian volunteers and drunk pretenders were replaced by trained professionals. In 1939, smokejumpers joined the cause, a fleet of daredevils most easily described as the Navy SEALs of wildfire fighting. When a fire lit up, they were sent in planes to parachute down to the remote flames and begin building Harry's lines. Seven years later another courageous force, the first hotshot crew, was created. Those were firefighters who trekked in by foot to fight flames. The work was treacherous of course, but thanks to Harry's weather predictions, trained teams could take measured risks instead of shots in the dark.

But 1949 showed the science of taming fire still had a long way to go. Thirteen firefighters, including twelve smokejumpers, were killed on August 5 of that year while trying to outrun what's known as the Mann Gulch Fire, which devoured 3,000 acres in ten minutes. Wagner Dodge, the leader of the group, saved himself with a last-minute decision to build a kind of escape route using flames, something that had never been done before. As the Mann Gulch Fire approached, he lit a small fire, hoping it would quickly burn out enough dry vegetation to ride out the passing firestorm. All he had left to do was lay down in his patch of freshly charred trench and pray. The firestorm passed and he miraculously emerged from the ashes alive.

Months later, Harry hiked to the gulch to investigate the explosive fire and document the pioneering method Wagner used to survive the flames. He had never seen fire behavior like this before, he told a colleague who had joined him in the field. Terrain, he realized, played yet another role in a fire's rate of growth. When wind was blocked by obstacles like tall rock formations, hills, or mountains, it spun off the sides more energized, not unlike how rocks in a stream create rapids after water flows around them. To be downstream, or in this case downwind of a moving fire, could be deadly. Harry was looking forward to comparing his findings to his other notes back in the lab. He then sat down on a rock, mentioned his legs were achy from a long day of hiking, and died of a heart attack. He's often referred to as the fourteenth fatality in the Mann Gulch Fire. He was only fifty-six years old at the time.

Harry's research helped develop a roster of techniques that are still used by firefighters today. He grew the West's reserve of wildfire firefighters from zero to more than 50,000 men and women working on the federal, state, and local level, and now we benefit from crews trained to employ techniques Harry developed a century ago without the assistance of aircraft, radar, satellite imagery, or any other modern convenience now taken for granted in the effort to tame our climate.

Harry's U.S. Forest Service was one of the greatest examples of eco-engineering in the world at the time. And for a while, it worked. The 1900s were marked by relative fire stability. But holding back Nature's fire cycle came at an unforeseen cost. Harry's research didn't factor in population growth and its compounding side effects: climate change and habitat change. As more and more humans spewed more and more heat-trapping greenhouse gasses into the atmosphere, making the planet hotter than ever before, we also weakened Mother Nature's ability to fight the fever by chopping down her trees, redirecting her riverways, hunting her wildlife (our planet's eco-engineers) to near extinction, and sucking more and more water from her already arid soil. The early signs of ecosystem collapse were setting in. Human expansion—from cities to towns to woodland neighborhoods—made it even harder, and in some cases impossible, for firefighters to prevent wildfires and access them once they started. Putting out fires by ten the next morning? Yeah, right.

Soon, even Harry's army wouldn't be able to hold back the firestorms on the horizon. The end of the twentieth century and beginning of the twenty-first gave birth to a patchwork game plan for fighting fires. Some blazes would be put out immediately. Others would be allowed to burn when possible. It was the imperfect science of man losing his short-lived power over the elements.

Harry's work still continues inside the Missoula Fire Sciences Laboratory, the place where his portrait hangs. Their findings could help revise that dated playbook so that man and nature can coexist in our rapidly changing ecosystems. But even twenty-first-century science is struggling to decipher the smoke signals of today's megaflames.

Nature Fights Back

I arrived at the scene of my first wildfire wearing boat shoes. Twelve hours earlier I had been settling into a lobster dinner with my family and quietly mourning the last evening of vacation on Cape Cod when my London bureau chief Andy Clarke called with a favor: "The West Coast bureau needs help covering a wildfire. If you leave now, can you get to Los Angeles in time for the morning show?"

The timing seemed impossible, and I could have said no, but you already know how this part of the story ends. Maybe what you don't know is the collateral damage unrestrained ambition and curiosity can cause, because after I hung up with Andy, I spun around the house stuffing shorts, shirts, and everything else scattered around my bedroom into my suitcase. My husband, Ivan, helped get me an Uber. My mother, ever the worrier, was certain there wouldn't be one available to pick me up. "There's going to be a lot of beach traffic on Route 6." My father, sensing my rising guilt for agreeing to go, eased my concerns: "Don't worry about it, Jon. Either way you'd be gone by tomorrow."

My younger brother, Greg, rushed as well, busy arranging his camera for the now emergent family portrait we always take on our last evening together as my older brother, Marc, and his wife, Jessie, corralled my nephew Leo into clean clothes for the abrupt Kodak moment. The wheels of my Uber crackled over my parents' crushed-shell driveway just as we finished taking the last photo. I never even had a chance to eat my lobster.

CBS News had a driver waiting for me as my flight skittered from Boston Logan International Airport to LAX, landing well after midnight. His black Escalade seemed better suited to a red carpet than a firestorm, but he said he was up for the uncertainty down the road. Where that road led us, after a winding 60 miles south through a rugged and parched hillside, was to rows of newly built homes with sprawling emerald front yards. The lush grass behind white picket fences was meticulously landscaped and dewy from sprinkler systems spritzing through the air in advance of daybreak. Our headlights glistened off each mirrored drop. It was a suburban oasis in what might as well have been a desert; a mirage that was actually very common in California, according to my driver. Thousands of similar neighborhoods dotted the state. They were so common, there was a technical term for them. The "wildland-urban interface"—a zone of transition between wilderness and land developed by humans. On this day, this built environment I was traveling through was about to have a standoff with the natural one it was trying to domesticate.

The driver passed through several police blockades before arriving at a cul-de-sac where local and national news teams were staged along with engine crews who had been monitoring the flames. The bright orange sky pierced through the windows of the homes when we pulled up, making them look like jack-o'-lanterns. The campfire scent in the air told me there was an inferno nearby, but I couldn't spot the flames and the few firefighters I did see were sleeping in their trucks and sprawled out on driveways. Some were even passed out on the wet grass. The threat was clearly far enough in the distance, but my deadline was looming. I had fifteen minutes to spare before going live.

"Where are your fire boots?" asked my CBS News producer Simon Bouie. He had bright brown eyes framed by tortoiseshell glasses and was wearing a buttoned-up plaid dress shirt. It was our first meeting. "You're wearing them," I shouted back before breaking the predawn tension with a laugh. "But seriously, do you have an extra pair?" Simon brought me into the satellite truck—a tank of a vehicle that's responsible for beaming live video from the field to a control room in New York that then feeds it to millions of televisions across the country. I quickly tracked—or voiced—the story that had been written by producers in the L.A. bureau, and after brief introductions with the photographer, audio technician, and truck operator, was escorted by flashlight to the live shot location in the backyard of a threatened home.

From there I could see in the distance the flames that tinted the sky. With less than two minutes to go, I put on my mic and pressed an "interrupt for broadcast," or IFB, into my ear. The audio device, used by club bouncers, secret service agents, and the like, allows correspondents in the field to hear the live broadcast and, with the help of a separate microphone, communicate with both the control room and anchor back in New York. "How are the conditions around you this morning, Jonathan?" Gayle King asked a minute later live on television. I described the strong gusts of wind and how they were fanning towering flames in a remote canyon, causing an explosion of embers that glittered the sky. "The concern," I explained, "is those embers, carried by the wind, could ignite homes like the one I'm standing next to." When my report was done, I took a moment to look back on the fire and remember thinking, with a tinge of disappointment, it was at least half a mile away and would never reach the neighborhood I was in. I had no desire to see someone lose their home, but I did want a closer look at the firefight underway. I walked back to the live truck in my three-sizes-too-big, borrowed boots, wishing I was still on Cape Cod with my family.

The rest of that morning was filled with proper introductions over breakfast at a nearby Denny's and planning what we'd film for the evening news if the fire continued to burn in remote terrain. Simon briefed me on typical wildfire coverage over eggs sunny-side up and coffee with cream and sugar. His manner of articulate speech, like his thorough notes, reminded me of a university professor. "Wildfires burn in relatively predictable patterns which make them easier for fire crews to control." As he explained, crews used bulldozers to dig control lines so flames would be cut off from dry vegetation and couldn't cross over to homes. "If that doesn't work, DC-10 Air Tankers are brought in to drop fire retardant. It's neon pink, which reads really well on camera," he said. He was essentially describing how Harry Gisborne's modern army worked, and according to this time-tested system of fire suppression, the flames would quickly burn through the man-made corridor and eventually come to a halt, choked by the fire lines. The results were usually easy to film, made-for-TV events. Guts and glory with little real risk.

But California's increasingly explosive and unpredictable wildfires were becoming harder to tame with traditional techniques, and coverage of them was starting to shift as more firefighters and news crews were forced to retreat from the front lines. "Not every correspondent is eager to cover fires these days," Simon said as we settled our bill and continued with our gathering for the day. "I can't believe you volunteered for this," he laughed. It was August 2018 and people were just beginning to pay attention to the monsters Harry first chased a hundred years earlier. Even he would have been surprised by the new class of fire that exploded in the twenty-first century. The 2017 Tubbs Fire jumped over a six-lane highway, a formidable control line, and ripped through 36,000 acres in Napa and Sonoma counties. It killed twenty-two people and destroyed more than 5,000 homes, businesses, and other structures according to the California Department of Forestry and Fire Protection, or CAL FIRE. The agency employees around 2,000 firefighters statewide who respond to everything from wildfires and floods to search and rescue and earthquakes (this is in addition to thousands of other local, state, and federal crews that are often called in to assist). At the time, Tubbs was the deadliest blaze CAL FIRE had ever battled. Less than a year later, and only a few weeks before I arrived in California, the Carr Fire tore through several neighborhoods in Shasta County, killing eight people, including three firefighters. Deadly fires burning down towns was incredibly rare. Twice in less than twelve months? Unheard of.

"The West has never seen anything quite like this before," said Mike

Pechner, a meteorologist who monitors wildfire weather as a consultant for CBS News. I spoke with him on the phone that afternoon as
he helped guide our field team to where we might find flames. "The
severe heat and drought are really fueling these things," he said. "Imagine having a 103-degree fever and being wrapped in a heavy blanket
you can't take off." That blanket was heat-trapping greenhouse gas,
the driving force behind our changing climate. California's average
summer temperature had risen more than 3 degrees since the turn of
the twentieth century, with more than half of that happening since
1970, according to the Scripps Institution of Oceanography. That's
enough to fuel a historic yearslong drought that made vegetation drier
than paper and helped kill millions of the state's trees. "There's more
water in the lumber at the Home Depot than in some of these trees,"
a firefighter told me. "And what's really complicating all of this is we
don't have the resources to clear all this vegetation. We've never seen
these kinds of fuel loads before," Mike said.

Traditionally there are two kinds of wildfires. Brush fires explode
quickly but are short-lived, and forest fires creep slowly but burn longer. If you've ever built a campfire, then you instinctively know how
it all works and how climate change and habitat change are turning
the kind of fires you heat s'mores over into raging bonfires. The key to
any good campfire is a combination of tinder—things like paper or dry
grass, bigger pieces of kindling and heartier logs on top. A campfire of
only newspaper would explode but just as quickly smolder out, similar
to brushfires like the one I was covering in Lake Elsinore. A campfire
of every ingredient is more stable in its growth and more destructive
as it reaches higher temperatures, like a forest fire. But with higher
air temperatures in the state creating more fuel (climate change), and
decades of fire suppression preventing the natural removal of it (habitat
change), a study by the University of Maryland's Global Land Analysis
and Discovery laboratory found both brush and forest fires are burning twice as much land and hotter than they did in the early 2000s,
and in unpredictable ways. Humans have essentially thrown kerosene
on today's flames. There are also more of these massive blazes than

ever before. The number of large fires between 1984 and 2015 in the western United States has more than doubled, according to NOAA.

My team and I spent the better part of the day looking for access points that would get us closer to the firefight, but found nothing. Mike warned us to avoid valleys, which could act like wind tunnels and cause rapid burning. "I doubt even a wide fire line will be able to hold back the fire if there's a good gust of wind," he cautioned. We ended up back where we began ten hours earlier, behind the home where I first spoke to Gayle. The fire had burned much closer but had been contained to hotspots. We had thirty minutes to go before the *CBS Evening News* broadcast and I was aware of how little ground the fire and my reporting had covered that day.

Suddenly, the air started to swirl, blowing embers like dandelion seeds. Hotspot after hotspot ignited closer to homes, and, fanned by the increasingly stronger gusts of wind, began forming what were later classified as fire tornados several stories tall. In a matter of minutes the quiet backyard I'd prepared to go live from resembled a war zone. The front line had come to us.

Helicopters were now circling nearby, and then a "spotter plane" passed overhead followed by the DC-10 Tanker it was guiding to the flames. With little warning the massive aircraft swooped in and emptied its belly full of that photogenic neon-pink flame retardant. The tanker was so low, the power released from its three turbofan GE engines quaked the ground I was standing on and left the air ringing in its contrails. The fire recoiled, and I turned back to the job I'd been sent there to do. This entire sequence had unfolded just yards behind me as I reported for the *CBS Evening News*. "This is exactly what crews were afraid of," I explained, ". . . these flames making their way over this hill so rapidly, directly in the line of sight of these homes. It's important to point out at this hour, 20,000 homes have been evacuated for this exact reason." Steven Spielberg couldn't have directed it any better. "I've never seen anything like that before on television!" my executive producer at the time, Mosheh Oinounou, told me after the report.

The home I was in front of was ultimately saved with some partial damage, but the fire destroyed eighteen others before crews were able to get the upper hand. What stood out most to me, as I later reflected in my hotel room, was how little control even the best trained and equipped teams seemed to have over the flames. As wildfires go, fighting this brush blaze was supposed to be a piece of cake. I returned to London three days later with a reverence for fire and a deep-seated fear of what the worsening side effects of climate and habitat change would mean for an ecosystem already burning out of control.

A week later I received a call from CBS News's head of talent, Laurie Orlando. "How would you feel about moving to Los Angeles?" As a journalist craving to cover our environment, I couldn't think of a better front line than the American West. We set a move date for early the following year—naïve, I suppose, in our expectation that Mother Nature would wait for my arrival.

Paradise in the Crosshairs

The towering welcome sign leading into Paradise, the one people driving from Chico passed by, looked like a page out of a family scrapbook—the kind with photos and keepsakes glued into a collage. The dark brown wooden structure, about 25 feet tall, was layered in colorful plaques of various shapes—the logos of the thirty-four different businesses and organizations that sponsored it. There was the Boys & Girls Clubs of the North Valley, the Ridge Coalition for Peace and Justice, the California Veterans Square Dance Association, the Quilters Guild, and the Men's Garden Club, to name a few. Just looking at that sign, you got the impression the town was loved, and those who traveled beyond it for the first time were always pleasantly surprised to find a community that lived up to the endorsement.

To locals, Paradise was known affectionately as the "The Ridge," a nickname that caught on "no one remembers when but stuck ever since." It was the kind of place you see in the kind of movies where a guy takes his girlfriend home to meet the parents: cue quirky shop owner, charmingly nosey neighbor, and nostalgic dive bar full of old friends from school. There was something for everyone here, and some things everyone liked, like Aunt Mabel's General Store and the Black Bear Diner that was always packed, especially after high school football games. The Ridge was deceptively quaint. And I say "deceptively" because 30,000 people isn't exactly quaint. For scale, that's around the size of Beverley Hills, California; Fairbanks, Alaska; Bangor, Maine; and Ithaca, New York. Now imagine erasing one of them from the map. Homes, schools, churches, a hospital, restaurants, supermarkets, a coffee roaster, landscapers, a veterinary office, hotels, garden centers, daycares, fast-food joints, an auto body shop, clothing stores, a funeral parlor—one day they're here, the next day they're gone. On November 8, 2018, that's exactly what happened. A total of 18,804 businesses and homes lost, thousands more damaged, and eighty-five people dead in what became at the time the country's most destructive, and California's deadliest, wildfire in a century. Survivors remember passing the welcome sign as they escaped that autumn morning. It, too, burned down, leaving only the stone slab it rose from.

I never told you what the message on that welcome sign said, did I?

MAY YOU FIND *PARADISE* TO BE ALL ITS NAME IMPLIES.

It was written in bold gold lettering. In the months that followed the fire, a new, unendorsed sign was propped up.

LOOTERS BEWARE. WE HAVE GUNS AND A BACKHOE. KEEP OUT!

Nature's Hydrogen Bomb

Northern California's Diablo Winds earned their name—Devil—for their dry and hot nature as they whistle through the Sierra Mountains. On the morning of November 8, 2018, you could see the Devil in the shape of the autumn leaves as they somersaulted along mountain roads, and in the twisting, bending, and cracking of oak branches that collected on the ground where they trapped the road-rashed foliage. Before long Diablo had built thousands of little tinder-and-kindling bird's nests. The only thing missing now was a match. Time would fix that. According to results from good old Harry's fire danger rating system, still going strong after all these years, the chance of a blaze sparking and growing rapidly was 76 percent. The vegetation was the driest it had been in years and wind gusts were already reaching 52 mph. Soon, trees shook violently from left to right, front to back. Dislodged acorns bounced in every direction. At one point Diablo blew so quickly and in so many different directions at the same time that leaves and dust appeared suspended, making the forest look submerged underwater.

The electric transmission line on Camp Creek Road was built to withstand hurricane-force winds, but that was nearly a century ago. The aging equipment, managed by the California utility company Pacific Gas and Electric (PG&E), had become threadbare with time. Diablo, now blowing 70 miles an hour, knocked out the old line at 6:15 a.m., creating a spark that fell onto those bird's nests below.

Residents in the town of Paradise remember waking up to the sound of the raspy howl of wind. The branches of trees and bushes were scraping on windows and roofs and siding. Those who looked out their window could see an ominous red glow in the distance. Residents of Paradise had developed a conditioned complacency to fire, thanks to firefighters' sterling record of putting previous fires out before they threatened people or property. But something about today was different. According to a timeline provided by CAL FIRE, residents—worried about smoke and that red glow on the horizon—began calling 911 around seven a.m., as school buses made their rounds to shepherd children to their morning classes. Soon the emergency line was flooded with hundreds of similar reports, all of them answered by the only dispatcher on duty before backup arrived. At the time, the blaze was ten miles away—too far to pose any real threat, Paradise officials thought. In audio recordings released by CAL FIRE, that operator eased concerns. "I've got to find out where this fire is in Paradise. I have children in school," one caller said. "It's not in Paradise, ma'am." "So far we're not in any danger," she told another caller. "This looks so close. I can see that it's orange," a woman's voice trembled with fear. "Ma'am, unless you see flames, then it's just the sun shining through it." The dispatcher even advised one resident to "stay where you are and close all your windows" to prevent smoke from getting in the house. She had the soothing confidence of a doctor instructing a patient to take two aspirin and go to bed. That advice certainly was reassuring, and would have been right had this fire been like all the others. What happened next, though, defied all wildfire modeling systems created to predict fire behavior.

Earlier I wrote that traditional forest fires creep in slow and predictable patterns on the forest floor, presenting the opportunity to develop a standard operating procedure before their arrival. But decades of climate and habitat change were fueling a fire that was anything but traditional. In short order the Camp Fire became airborne as Diablo blew embers miles away, creating new blazes. The wind was so strong, flames jumped from treetop to treetop, with heat from the exploding blaze radiating so torridly that it changed the genetics of the atmosphere. Meteorologists would later say the Camp Fire generated its own weather system. Diablo had created a monster it could no longer control. The blaze was now devouring eighty football fields a minute, ripping through ten miles in less than two hours, according to CAL FIRE! The flames reached the smaller town of Concow first, which, like Paradise, hadn't been evacuated. Scanner traffic captured the chaos. Residents were trapped. Some had jumped into a lake. A group of firefighters deployed their heat-reflecting metallic fire shelters to survive.

That's when Paradise officials, hearing that scanner traffic, began to panic. The first evacuation order came at eight a.m., but it was too late. By the time the alert rang out, 911 was getting calls of flames at the eastern edge of town, according to CAL FIRE. A dozen spot fires quickly turned into a hundred. Thick acrid smoke blocked out the sun. Aunt Mabel's was on fire. Smoke was seen coming from the Black Bear Diner. Evacuations were issued zone by zone to prevent overcrowding on the roads using CodeRed, a new phone alert system that required residents to voluntarily submit their phone numbers. "There's an immediate evacuation order for zones 2, 3, 7, 8, 13, and 14 due to a fire. You should evacuate immediately," one alert read. But according to Paradise's Emergency Operation Center, only an estimated 25 to 30 percent of the town had enrolled ahead of the Camp Fire, and even some of those who did sign up still didn't get a notification. The Paradise Police Department would later reveal as many as 60 percent of CodeRed's alerts failed to go through. While evacuation orders were slow and even silent, panic spread by word of mouth at rapid speed, and in short order people from every zone flooded the roads all at once.

CAL FIRE had warned Paradise officials back as early as 2005 that the town's infrastructure could no longer accommodate its swelling population. Skyway, the only main road out, was not wide enough to handle an evacuation of the town's 28,000 residents in the event of an emergency. Paradise's local roads were also narrow, old orchard paths that had been paved over. Navigating them was no easy task on a clear day, let alone in a smoky and fiery daytime twilight. It didn't help that more than a hundred miles of these narrow paths led to dead ends. CAL FIRE also expressed concern over the buildup of dry vegetation and dead trees—killed by drought and the state's bark beetle infestation—that were left to collect on the floor. The forest had decayed into a powder keg. Then there was the terrain. Paradise sat in the middle of several wind tunnels, like that proverbial rock in a river.

A *Los Angeles Times* investigation found that Paradise officials "ignored repeated warnings of the risk its residents faced" and "crafted no plan to evacuate the area all at once . . . It remained doomed because for all the preparations community leaders made, they practiced for tamer wildfires that frequently burned to the edge of town and stopped—not a wind-driven ember storm." A local firefighter told me that historically, no fire had jumped over the nearby Feather Creek, which separated the town from the wild, overgrown fringes. Paradise officials soothed themselves with the idea that they had time to make the costly updates to its infrastructure, not realizing the rapidly changing climate and habitat made historical data less reliable and every second borrowed.

As the Camp Fire jumped Feather Creek and ripped through town, the bottleneck out of Paradise saw Skyway choked by a miles-long gridlock. Meanwhile, the journey to safety hadn't even begun at the Paradise elementary school. Teachers didn't know what to do. When they were notified of the evacuation, they tried calling parents, but many calls weren't going through. Teacher Katy Schrum stayed calm and started packing kids into the same school buses that had dropped them off an hour earlier. "We were running out of time, and we made the decision right then and there that we needed to get them to safety." Among the students was seven-year-old Ellie Wrobel. She was about to get on the bus when her mother, Kylie, pulled up to get her. "Where's Daisy?" Ellie asked. The family's black Lab was still at home. Kylie had come to school straight from work, she explained. The fear that glazed their eyes reflected a worst-case scenario. Today was more than just a drill, and there was no way they were leaving Daisy behind. The cabin of their car was quiet as the two of them stared out the window. The bus Ellie had been about to board was on the rearview road to asylum. They were the only car trying to get back into town.

That day it took people five hours to drive just sixteen miles to Chico, and geography ultimately separated the winners from the losers. People on the western side of town, the side closest to the finish line of this mice maze, escaped physically uninjured, though many couldn't stop shaking for weeks because of frayed nerves and night tremors. Those on the eastern side of town, where Kylie and Ellie lived, fared far worse. The fire burned strongest here, and many got disoriented in the smoke. Several people drove down roads, escaping fire behind them, only to be stopped at fiery dead ends where they were swallowed by flames. Others lost sight of the road altogether and drove into ditches, where they were trapped. The fire burned so intensely that those stuck in traffic struggled to breathe as the flames consumed all the oxygen around them. Brain hypoxia set in, with symptoms including memory loss, reduced mobility, and difficulty paying attention and making sound decisions. That's the only way to explain why some thought it best to abandon their cars and escape on foot, blocking dozens of other drivers in vehicles behind them. There was no way to clear the blockade before the flames arrived.

In one of the most haunting videos to emerge from the tragedy, a man filmed the burned-out wreckage of a sedan with the skeletal remains of his neighbor still behind the steering wheel. "My friend! You can see he's dead. And so is his mother," he said through tears. "Rest in peace." Then there was video that offered hope after dozens of people survived by riding out the firestorm in the concrete parking lots of businesses. Firefighters who had been helping direct traffic hatched the plan when it became clear they had run out of time. Embers exploded all around like the death of a thousand stars.

I watched this nightmare unfold on television from my office in London and thought about my time in Lake Elsinore just a few months earlier. Fire crews there spent nearly a day building control lines. The

Camp Fire was moving too fast and behaving far too erratically to be corralled. Firefighters never had a chance. Many never even had time to lift a hose as they helped people escape.

"Let's all just be honest," Jim Broshears, the director of Paradise's Emergency Operations Center, later told reporters. "We had an evacuation plan built for a wildland fire. We got a hydrogen bomb instead."

Picking Up the Pieces

"Book an early flight into Chico tomorrow and grab a rental car," the email began. "It's about a 40-minute drive. Don't forget a face mask. It's still very smoky." I had only just begun unpacking the cardboard boxes in my West Hollywood condo when I was assigned a story about Paradise's recovery. It was early May 2019, six months after the disaster. It took just under five hours to make my way from congested Los Angeles to the folded crease of Northern California's map. I pulled into a gas station before arriving in town to get a Red Bull and a stick of jerky. At checkout the clerk asked what news agency I was with. He must have seen the press pass in my wallet or picked up on a scent I'm always worried about giving off. That unique musk of a journalist who promises more than they can ever deliver.

"CBS News," I said. "We're covering the rebuilding."

"Good," he said. "Everything else has been fire porn."

I understood the sentiment. My arrival in Paradise coincided with the invisible expiration date a community stamps on the press when we first start knocking after a disaster. In a tragedy's immediate aftermath, journalists are a welcomed sight. Misery doesn't just *love* company, it *needs* it, like we all do when we're in trouble and a rational friend steps in to make sense of what the fuck just happened. In those early days, the press can play a critical role in helping communities process trauma and navigate through the red tape that often slows down the aid required to recover. But there is a fine line between public service and self-service. Eventually journalists start to curdle, and the worst of us turn into feral alley cats prowling for scraps.

On this day, my team and I had been invited. Paradise still had a long way to go before it was whole again, and feral or not, help was help. There was no line separating Paradise and its fate from the untouched communities around it. Instead, the city limits were drawn in cinder. I'd been informed the fire had been devastating, but there was no way to prepare for what I was seeing now. The only colors left amid the devastation I was driving through were reflective yellow and orange vests of the cleanup crews that lined every street for miles. Teams with the Federal Emergency Management Agency (FEMA) were on site assisting. Ninety-five percent of the city was erased, and nearly four million tons of wreckage needed to be removed, according to the Governor's Office of Emergency Services. That's double the destruction caused when the twin towers collapsed on September 11, 2001. In a single afternoon, the equivalent of four 110-story towers fell in the forest. The Skyway Antique Mall was among the few commercial structures that survived. The beige concrete-box-of-a-building looked like a tombstone memorializing everything around it that didn't survive. Before the Camp Fire, you could fly over Paradise and not even know it was there, hidden under a canopy of trees. On this day their charred trunks were being hauled off on massive flatbed trucks.

In its current condition, Paradise was no place for a mother and daughter to live, but fire and brimstone molded Kylie and Ellie Wrobel into modern American pioneers. They had lived on one of those dead-end orchard roads, and perhaps because they knew this confusing web of capillaries—many not even drawn on maps—like the back of their hand, they beat the odds and escaped with only emotional scars. Their home—and everything in it—was destroyed, like thousands of others. Paradise Elementary, Ellie's second home, was lost, too. They were now living in a loaner RV on the outskirts of town, among just a handful of residents who had returned to help rebuild.

I saw Kylie and Ellie before they saw me. They were sitting in chairs side by side, speaking softly to each other, both dressed in black hooded sweatshirts with blond hair parted down the middle. Kylie's was long and straight with highlights, Ellie's short with Shirley Temple curls that sprung up and down against her pink-framed glasses and puckered face. Ellie had recently turned eight, but had developed the perceptive squint of someone who had seen many more trips around the sun. The fire was the kind that turned kids into adults and rattled parents with childlike fear. On this morning they looked more like sisters than mother and child, the frayed ends of the lifeline between them pulled closer by the trauma they'd seen. I introduced myself and thanked them for taking the time to share their story. The cameras rolled and our interview began.

"Why did you decide to come back to Paradise?" I asked.

"There's nowhere else for us to go. Nowhere else that we want to go. This is home. Paradise is our home," Kylie said in a soft monotone.

Kylie was born and raised in Paradise, just like her mother and father before her and her grandmother and grandfather before them. All people rooted in forests share a connection to nature those born elsewhere don't, and in this town the knots and limbs and flittering leaves of every mature conifer, deciduous, and evergreen were land-marks of life. People spoke of the oldest giants with reverence befitting an ancestor. While I'd arrived seeing a pit of ashen devastation, Kylie saw the smoldering remains of her family tree. "I know a lot of people aren't ready to come back and probably never will. But this town is part of who I am. I believe what's happening with our climate is real, but I also believe we can prevent another disaster."

As I sat there, my knees a few feet away from theirs, I was amazed at how strong she and her daughter were, especially Ellie. Kids are resilient, I thought. But then those aged eyes, staring directly into my soul, started to water, and soon her lids were unable to hold back tears.

Kylie, who was surprised by Ellie's emotion, began crying, too, as months of blistering trauma ruptured. They had both been suffering in silence to protect each other. "She does bottle up stuff and it's a good release for her to talk about it. We've tried to focus on the positive and the future, but we have a long way to go before we can fully heal," Kylie said. "What I'm learning is there's not one right way to move on from something like this. We're figuring it out day by day." The next hurdle was building a new home. "Six more months was the latest update we got from the contractor," Kylie told me. I promised I would come back as soon as possible, but Paradise, it turned out, wasn't a fluke, and I spent the next two years reporting on other Western towns with similar fates before returning.

Engineered Miracles

The plastic ring dangled from a white rope overhead. It was just out of arm's reach, and it took me a few days to even notice it swinging from the ceiling in the air-conditioned breeze that flowed through the foyer. The rectangular hatch it was attached to, disguised to blend in with the crown molding, wasn't advertised by the landlord of the West Hollywood condo my husband and I rented when we first moved to Los Angeles. Using the narrow-end of a wire hanger, you could lasso that plastic ring and pull down a folding ladder that took the curious up to a cobwebbed attic. Here, alongside a pack of light bulbs, a rusty can of Sherwin-Williams touch-up paint, and a crusty red mop, was where I kept my wildfire go-bag, which smelled of barbeque from covering previous fires.

Go-bags, as their name suggests, are prepacked field kits with essentials needed for last-minute assignments when packing is a luxury breaking news can't afford. I have several versions of them depending on the story and season. My wildfire kit, a dark blue duffel, is stuffed with things like a pair of bright yellow pants and matching jacket, both woven using fire-resistant material called Nomex. I also have a helmet, smoke mask, goggles, and fire boots along with a two-way radio and a pop-up fire shelter—an aluminum-foil sleeping bag that deflects radiant heat, up to 2,000 degrees Fahrenheit, for about a minute in the event a person can't outrun the flames. It's a piece of equipment firefighters hope they never have to use, and most never do. I guess that's why I developed the bad habit of removing mine before grabbing my bag and going. If firefighters rarely used theirs, then I would never need mine, I thought. After all, I wasn't a trailblazer battling flames. I just followed far behind, shielded from danger by *their* sacrifice. Plus, ditching my shelter helped lighten the load. Each one weighs about five pounds.

It should come as no surprise that a firefighter's list of gear is much longer than mine. Firefighters trek to the front lines wearing up to seventy-pound backpacks, with everything from line gear and hose packs to first-aid kits and drip torches. Some carry chainsaws. Most carry Pulaskis, a type of axe that doubles as a hoe. To stand next to a fully suited wildland firefighter is to understand why children in the mountain towns they protect place those yellow uniforms, with patches from CAL FIRE or the U.S. Forest Service sewn to the shoulder, on a rotating list of Halloween costumes that also includes Superman and Captain America.

But to really study a firefighter's uniform and gear is to also understand these heroes are mere mortals. Between 2007 and 2016, the last time a tally was taken, 170 firefighters died battling western wildfires, according to the National Wildfire Coordinating Group. While that yellow uniform can protect fighters from the embers, heat, and small flames of Mother Nature's creeping fires, they are increasingly an expired insurance policy against bigger flames, where escaping, if surrounded, requires an act of magic. Which brings me back to those fire shelters. They've been a last resort more in recent years than ever before, and countless lives have risen—hot, often singed, but alive—from the ashes. Engineered miracles.

On the night of October 22, 2019, I grabbed a metal hanger, lassoed that dangling plastic ring, and climbed the ladder to grab my go-bag from the attic. I left my dusty fire shelter behind the night Napa caught fire. That night, something happened I'd never seen before. The kind of thing that breaks bad habits cold turkey. On that night, the ashen boot prints that marked the protected path I followed to see and report on fire would be used by its very creators to flee. I barely escaped what happened that night. While I credit faith for surviving, from that moment on I'd start packing my fire shelter—my own modest bag of tricks.

"Go, Go, Go"

October 22 began like any other Tuesday. I woke up and went to the gym, spent the day working on pitches for my next stories, and by four p.m. was shopping for dinner at the grocery store. I was in the dairy section when I got the call. Red flag warnings had been issued in Northern California, meaning high winds—the breath that blows fire—were all but certain. That coupled with a heat wave and yellowed land raised the fire danger to "extreme." "Can you base yourself in San Francisco in case something sparks?" my CBS News bureau chief at the time asked. If it wasn't clear yet, saying no is rarely an option in this line of work, and so I went home, grabbed my go-bag from the attic, and darted to LAX.

I remember thinking how foolish I must have looked running to the gate that evening, all so I could babysit Nature's whim from a hotel room near Fisherman's Wharf. It didn't help that when I finally arrived around eleven that night, the minibar was empty. I typed my restless mind into the notes app on my phone—"call Amex, pay parking ticket, submit vacation request, coffee @ Buena Vista"—and once cleared of thought, drifted off to sleep. An hour later, my San Francisco–based producer Chris Weicher was on the phone. "Wine country is on fire. Pack your stuff and Uber to my house so we can drive together," she said breathlessly. I was out the door in minutes.

The developing wind had pushed the fog out of San Francisco Bay, and the Golden Gate Bridge was lit up as bright as a full moon, revealing every nut and bolt and steel seam. There were no other cars in front of us, and the shadows that strobed through my Uber's sunroof made it feel like we were going faster than we really were. The adrenaline was starting to kick in. I met Chris at a gas station close to her home. If the fire we were about to encounter was truly wild, you wouldn't have guessed it by the car she was driving. Chris has worked in broadcast journalism since Watergate, around the same time her maroon-colored Mercedes rolled off the production line. You'd never know by looking at her—with her salt-and-pepper hair and gentle hazel eyes—that she had covered more wildfires than she could count.

Have you heard of the American artist Thomas Kinkade? His paintings of storybook cottages nestled next to rivers and waterfalls and lakes and oceans have a Disney-like quality and were plastered on everything from coffee mugs to calendars in the '90s. Given all of that, I had a hard time picturing the Kincade Fire, as this one had been named, as the hellscape Chris had described. Wildfires are given their names based on where they start, and early reports said this one had sparked after windfallen trees snapped power lines on John Kincade Road in the town of Geyserville, California. Thousands of people in Sonoma County were under evacuation orders and homes had already begun to fall to the fire, Chris told me during the drive over.

A neon-red sky now silhouetted Sonoma's famous vineyards, which looked like a rave of scarecrows. The whole scene came to life in fits of flashing red and blue as emergency vehicles moved into position. We had all arrived around the same time. Within minutes our cranky Mercedes was surrounded by a sea of sirens. In California, journalists loosely fall under the category of first responders, which granted us all access despite John Kincade Road having been blockaded by police. An officer looked puzzled as Chris manually rolled down the window of her vintage ride and handed him our press credentials. His eyes locked with mine in a brief telepathic moment as he waved us through the closure: "Good luck," I'd like to think he said. "I'm fucked, aren't I?" was my response.

We drove through a narrow and winding canyon road with a wall of brush and trees on both sides. The flames hadn't scorched this part—yet. It took about ten minutes for us to reach our team at the staging area near the top of the mountain. Photographer Gilbert Deiz and sound technician Bryan Baeta were filming engine crews when we arrived. The glow from the flames drowned out the emergency lights as we pulled in. From our perch above the valley I was surprised to see the fire was burning far in the distance, on a remote ridge.

Maybe I was still too drowsy to process what was unfolding around me, but in that moment I thought Chris had received bad intel and sent us to cover a false alarm. Many wildfires burn out or are put out before they grow into monsters, and unlike my first fire in Lake Elsinore, the closest homes here were miles away. The threat seemed minimal at best. The canyon wind didn't feel all that strong and the firefighters that had rushed to the scene were now waiting and watching what the flames would do. If my time in Lake Elsinore taught me anything, it could be hours before there was action, if any.

Wildfires were still relatively new to me at the time, but I was starting to pick up on things I hadn't noticed before. Like the glint of disappointment in a firefighter's eyes, matched in my own, as the possibility of a rushed trip wasted washed over us. Gilbert, himself a tall and stoic figure chiseled by news events like the L.A. riots, had a different kind of look as he fixed his eyes on the source of the glow. He was twice as old as many of the crews on the front lines and knew better than to underestimate Earth's elements.

It was three in the morning on the West Coast, which meant we had an hour before the show broadcasted live from New York. We were the lead, but what would I even report? Gilbert and I surveyed the area. "Let's start on the firefighters and all the flashing lights here, and then walk over to this wooden fence to show the threat in the distance," I suggested. That fence, we later learned, was the perimeter of a pasture. Soon the wind began to whistle through vegetation. It was a broken song at first that grew steady, like a teakettle ready to be taken off the stove.

I was heading to Chris's car to get a can of Red Bull I had picked up earlier at the gas station when Gilbert called me back. "Let's tape a debrief in case things get out of control and we have to leave," he said. The wind had grown strong enough to carry embers into the pasture, and firefighters were now responding to hotspots. In television terms, a debrief is a minute or two report where the correspondent explains what's happening around him. There's no highly produced story to toss to. Just the single scene, as it's unfolding in real time.

In the time it took me to reach Gilbert and put my microphone on, a small tree about ten feet from us was engulfed in flames. The fire I'd thought was far enough away was suddenly on our heels, and I could now make out what appeared to be cattle in the distance rustling awake by the crackling flames closing in. I could feel the heat on my back, devouring limbs and leaves, as we recorded our spot from the field. I wasn't reporting on a fire hours before or after its explosion. I was standing in the middle of its soon-to-be-spent accelerant.

The kettle was now jumping on the stove. Hurricane-force gusts turned flames into sky-scraping tornados. Embers landed around us like fiery arrows as the source flames advanced at rapid speed on the ground. The quiet pasture quickly transformed into a war zone as fire flared and cattle cried, unable to outrun the blaze. Within seconds we, along with first responders, were nearly surrounded by a wall of fire. The flames were at least 40 feet high and only a few yards from us. And the heat was just incredible.

Red hot embers glittered in the air, creating micro explosions upon impact with the road. It was sickeningly beautiful, this field of a thousand dying stars. The road we were on was a wide enough fire break to temporarily stall the blaze's onward march, but it was clear this mountain pass, our only way out, would not be able to hold back the inferno much longer. "Get out! GO, GO, GO!" one firefighter yelled to us and his team, but it was his eyes that were the loudest. "Leave now or you will die!" they said without saying a word.

Chris, who was parked in front of the fire trucks, got a head start. Bryan jumped in his van and I piled into Gilbert's Suburban. The flashing lights of a fire truck in front of us guided our way. There were other emergency vehicles behind us. I was comforted by our placement in the middle of this convoy as we made our way down the narrow path to safety. The strobing lights echoed my beating heart.

But those fiery arrows were still landing in front of us no matter how fast we drove, and the road was now surrounded by flames advancing in all directions and spun in a blender of fierce canyon winds. Then came the smoke. The most dangerous fires are the ones you don't even see, concealed by a blinding and toxic layer of black smoke created by the rapid burning of wood, metal, and brush. The emergency lights disappeared. I couldn't even see the hood of Gilbert's SUV as we tried to keep up.

We continued on our blindfolded journey, confident at first that the worst would soon be over. But as asphalt turned to dirt beneath us, we realized we had accidentally turned off the main road in the blacked-out fracas, and were no longer heading away from the fire but looping around back into it.

Neither of us said a word. We both knew what was happening and were afraid the panic in our voices would somehow conjure the flames. We continued to drive, not because we thought we were heading in the right direction, but because there was nowhere else to go. Smaller fires began to form on both sides of the vehicle and brush up against the windows now too hot to touch. Smoke began to fill the cabin. "You're doing great," I said to Gilbert, speaking in short sentences to hide my shaky voice, thinned by the dwindling supply of oxygen available inside the car. "There's got to be a way out coming up." I was now conjuring an escape route that didn't exist.

It took another minute that felt like an hour before we pulled up to a dead end at what I later learned was a small casino. The road, and our luck, had run out. I had never had a "life flashing before my eyes" moment before. That morning, at that dead end, in that hot and smoky Suburban as flames moved in, I did. I thought about my husband and my parents, who always texted me to be safe before going on an assignment. I thought about Paradise and the dead-end orchard roads that claimed so many lives, and that cell phone video of a skeleton still gripping the steering wheel.

"Okay, do we stay or do we go?" I asked out loud, hoping Gilbert or God would chime in. In wildfire training you're taught to assess risk when facing an emergency in the field, and the risks we now faced were clear. If we stayed, the flames could engulf the car. And even if they didn't reach us, the smoke would be suffocating. Low visibility meant there was no clear direction to run if we needed to abandon the car in search of clean air. If we turned around we would be confronted by even stronger flames than the ones we'd passed. Would our tires melt? We weren't sure. And what if a tree had fallen and blocked the path? It's hard to think logically while sitting in a boiling pot. Was brain hypoxia setting in? Were we about to make an irrational decision?

As the flames moved in around us we decided we had no choice but to go down the same fiery road we hoped we had escaped. The air was so hot it hurt to inhale. Every breath, once taken for granted, was blistered by heat. You of course know the ending, but even now as I reflect, our escape that morning makes no sense. The road was a river of fire. We had zero visibility. I was drenched in sweat. It was an escape guided by faith. God or a spirit or whatever pins stars in the sky must have been listening that morning.

We choked up when we finally caught up with the convoy near a vineyard in the much safer foothills. The conditions had become too dangerous even for them. The entire crew of firefighters we were with had retreated from the mountain. "I've been doing this for fifteen years and have never seen anything like it. It's hard to train for these conditions," one firefighter said.

Gilbert, who had covered hundreds of wildfires before this one, told me he had never been as terrified as he was that day. "There was a moment I really didn't think we were going to make it," he said with a nod to the sky that let me know he wasn't talking to just me. Our watery-eyed confessional was interrupted by a knock on the window. It was Chris. "You have twenty minutes to feed the debrief. New York is ready when you are." She had no idea what we'd just lived through.

I've told this story of my near-critical dance with the Kincade Fire dozens of times. It was life-changing and remains seared in my mind, rising up again each time I set out on another reporting mission in pursuit of our unfolding climate catastrophe. In return, I'm often asked: Why not just report from the studio? I've been thinking about that question a lot lately, as I parachute into and out of one natural disaster after another, from hurricanes and floods to fires and heat waves.

To answer it, let's briefly go back to the year 2000. At the turn of this century, as you'll read in chapters to come, there were already more than 150 years of scientific data linking human activity to our rapidly changing climate and habitat. In fact, by 2000, extreme weather events were already outpacing predictions. And yet a poll conducted then by the American Association for Public Opinion Research found only 5 percent of respondents, when asked "how much do you feel you know about global warming?" said "a lot." *Five percent*. While it's hard to attribute this lack of knowledge to any one source, other data from the time does offer some clues. For example, the government transparency group OpenSecrets, which analyzes publicly disclosed data to track money in politics, found the oil and gas industry spent around $60 million a year in the preceding decade successfully lobbying against climate change legislation. Meanwhile, as an unbelievable amount of money was tipping the scales of public perception over human-caused global warming, the news media remained largely silent in its coverage. According to research by the University of Colorado's Media and Climate Change Observatory, which analyzes the media's monthly coverage of climate, the nation's top seven television networks aired a total of twenty-six segments in January 2000.

This all changed in 2006 with former vice president Al Gore's documentary *An Inconvenient Truth*, which raised international awareness by connecting abstract science to real-time destruction. At the time, I had just graduated college, and thanks to Gore's *Truth*, was beginning my career as a journalist with a new sense of purpose. The documentary's images were impossible to ignore, and not just for me. "Before the film, climate change in the U.S. was largely a debate about the scientific proof and partisan politics of the issue, whereas afterwards it evolved into an issue of ethical concern," research by Georgetown University found when analyzing the film's public impact. That "ethical concern" appears to have led to a spike in the American news industry's climate coverage. In January 2007, following the release of the documentary, the same seven main television networks aired climate change segments a combined 286 times, a 1,000 percent increase from 2000. Today, the Media and Climate Change Observatory finds that in some months there are as many as 670 segments, a more than 2,000 percent increase. Most Americans know about climate change today, and according to the Pew Research Center, 54 percent now describe it as a "major threat."

This education has helped drive solution-based initiatives. Since 2000, renewable energy projects like wind and solar have spiked 90 percent according to the Center for Climate and Energy Solutions. Today, more than ten million homes in California are primarily powered by solar. In Texas, wind provides 28 percent of its annual energy, a 200 percent increase from 2011, according to state records.

Could this all be coincidence? I think not. Without reports in the field and the irrefutable images gathered there—images that would otherwise not be gathered at all—what happens in the Lake Elsinores, Paradises, and Healdsburgs of the world, and the impact climate and habitat change are having on these communities, would be written off by history as unrelated, singular events. And let's face it. These events are still written off by some. While there has been progress, it hasn't been fast enough. Journalists are often referred to as the Fourth Estate, a term that recognizes the press's explicit capacity of advocacy and implicit ability to frame issues. For too long, that estate was asleep at the wheel. Today's climate reporting is part of the lasting legacy of those who first helped expose the crisis.

Why not just report from the studio? Because without the images and facts gathered from the field, we'd still be stuck in the 5 percent.

The day after our narrow escape through the darkness, Gilbert and I returned to that same dirt road. On our muted ride, we passed the smoldering remains of a home. The stumps of trees still flickered red in the occasional gust of wind, like the end of a cigarette. As we pulled up to the turnoff we were forced to stop. A 30-foot tree had fallen, blocking what was our exit that morning.

The Kincade Fire was put out on November 6, 2019, two weeks after it started. No one was killed, but 374 homes and other structures were destroyed. At the time it was the largest wildfire to ever sweep through Sonoma County, burning nearly 80,000 acres. Everything was about to change.

Western Blitz

If Paradise was a warning flare for the West's rapidly weakening eco-system, the 2020 season was a five-alarm fire. In a period of six months, I pinballed across the state as flames ripped through 4,304,379 acres, causing $12 billion in damage and shattering every state record dating back more than a hundred years. The flames even threatened Paradise once again, and Kylie and Ellie and about one hundred other people who had returned to the town by that point were told to be prepared for possible evacuations if the fire got closer. But the warnings were not of much use, as the few who had returned to Paradise had, as Kylie put it, so little left to lose. In fact, only a few homes had been rebuilt, and Kylie and Ellie's was not one of them. The two were still living in their donated RV.

In August of that year, a rare heat wave gripped California's Bay Area. PG&E, the local utility conglomerate, had protectively cut power to tens of thousands of customers in anticipation of yet another flash up. The next inferno would be impossible to stop, state officials warned. The lightning rolled in around two in the morning. The Golden Gate Bridge spasmed in the flickering light as powerful bolts fractured the sky. One witness on a local radio station compared the relentless strikes to the Blitz, the German bombing campaign against London during World War II. The more than 2,500 recorded lightning strikes turned

San Francisco night into strobing day, with 200 strikes happening in less than thirty minutes. But one critical element was missing: rain.

Mother Nature's dry, fiery assault struck trees, homes, power lines, and grassy fields, and sparked the beginning of what at the time became the most destructive wildfire season in state history.

Paradise—what was left of it—was spared just as Kylie had glumly forecast, but by the time the 2020 fire season came to an end, two other mountain communities had joined in its destructed state. According to CAL FIRE, 33 people were killed and 11,116 homes, businesses, and other structures were destroyed by fires that were so dynamic, unpredictable, and fast growing that officials gave these blazes a new nickname during press conferences that year: "megafire."

The more than four million acres that burned was double the previous record set just the year before. The state's drought and rising temperatures had drained natural vegetation of almost all its moisture, in a volatile trend that has continued in the years since, according to the U.S. Forest Service's National Fuel Moisture Database. While everyone had hoped Paradise was an outlier, scientists marked 2020 as a tipping point, the year humans officially lost their grip on the West. It took less than two hundred years to dismantle an ecosystem that had thrived for millennia.

The Whack-A-Mole Defensive

Sequoiadendron giganteum, better known as California's giant sequoias, have a life span of up to 3,000 years, allowing them to have borne witness to the full scope of industrialization's ravaging of the natural environment it plunders for fuel. Growing to 300 feet tall with trunks 100 feet in circumference, sequoias' bark can swell to two feet in depth and, in some kind of prehistoric genetic wizardry, pumps tannin—a fire-retardant chemical—through its veins, providing some armor from the region's ever more destructive flames.

Their prehistoric defenses reflect their prehistoric origins, a past that saw ancestors stand alongside the dinosaurs. Today, giant sequoias are classified as the largest trees in the world, and all that girth is birthed from a cone the size of a lime, smaller than the typical-sized pine cone you might find on the ground each winter. But sequoias are unique in another way that has proven ironic in our crumbling modern environment: The giant sequoia is born in fire.

Fire not only helps crack open the sequoia's cone, ejecting the seeds inside; it also clears the earth of any debris that could interrupt those seeds from burrowing into the ground and reaching Mother Nature's fertile loamy subsoil. Fire in forests is critical for plant reproduction. And not just for sequoias, but the entire woodland ecosystem. Fire clears dead vegetation, including invasive species of plants and bugs, which in return makes way for new growth—kind of like a lizard shedding its skin. And the nutrients from this charred vegetation then fertilize the soil, helping start the growth cycle all over again.

For tens of thousands of years, the Yurok, Karuk, and Hoopa tribes, native to what are now California's forests, recognized the vital role fire played in maintaining this natural balance. The Indigenous groups lived along streams inside wooden plank huts that sheltered extensive families. In Yurok culture, each hut was owned by the patriarch, and several generations of his bloodline lived under a single roof . . . sometimes as many as twenty to thirty people. Everything they owned and consumed came directly from the land. They fished for salmon in the rivers, sifted for shellfish along the coast, hunted game, and gathered plants on the forest floor. Because they lived off the land—and the seeds and nuts and grasses and animals it nourished—they knew it was in their best interest to learn to live with fire, not snuff it out.

Mimicking Mother Nature's own cycle, tribes even intentionally set fires in what Frank Kanawha Lake, a Yurok and Karuk descendant and research ecologist with the U.S. Forest Service, described to me .as medicine. "When you prescribe it, you're getting the right dose to maintain the balance and health of the ecosystem." As Frank explained, these blazes allowed Nature to run its course while letting California's Indigenous people live out their own lives in the woods largely undisrupted. During this period, up to 12 percent of the state was sprinkled annually with many small "low-intensity fires" sparked by lightning or set by Indigenous tribes, according to Frank's research for the Forest Service, which he wrote about in his paper "Historical and Cultural Fires, Tribal Managements and Research Issues in Northern California: Trails, Fires and Tribulations." It was not unusual for millions of acres to slowly burn over the course of a season. And because these flames were a regular occurrence and reduced fuel loads, rarely did a fire grow out of control. Ancient fire didn't destroy communities. It made them healthier.

Ecological records and oral Indigenous history show some tribes also saw land management by fire as a spiritual practice, even chanting "Wawona" during burns to, according to the National Park Service (NPS), represent the hoot of an owl, which was considered to be the guardian spirit of the trees. Over time, according to some, Wawona became the nickname for these towering shadow-casters. But the name white men would later give the world's single largest tree, which still stands in Sequoia National Park, tells the story of the colonization that changed the natural rules and sent a bountiful land of giants into a tailspin.

The cracks in this ecosystem first formed in the mid-1800s after the United States invaded and conquered California, then part of the republic of Mexico. The rapidly expanding United States was in desperate need of resources, and the West, rumor had it, was both a literal and figurative gold mine. Starting in 1848, what is now known as the California Genocide saw the U.S. government move in and force out the Yurok, Karuk, and Hoopa people. General William Sherman led this clearing mission, applying here a blueprint drafted and already carried out in the Midwest, where, in his words, the U.S. needed to "act with vindictive earnestness against the [Indigenous people], even to their extermination, men, women and children." A similar scorched-earth directive is exactly what unfolded under California's sequoia canopy.

In less than two decades, as many as 16,000 men, women, and children were killed and nearly 100,000 were displaced. California's woodlands were sanitized of all signs of its tribal past before the keys were handed over to white settlers, many of them unsuspecting of the horrific methods used to gain ownership—what today we would characterize as war crimes. In short order, small towns took root on this bloodstained forest floor and the U.S. Fire Service, with the aid of Harry Gisborne's research, implemented that poorly conceived ten a.m. policy to stop Mother Nature's natural fire cycle so humans could carry out their own. The engine ran smoothly for a while, and had our climate and habitat remained unchanged, perhaps today we would still be able to put out every firestorm by ten a.m. No such luck, though. Mother Nature was about to overtake man's defenses.

By the late '70s, as fires became harder to extinguish quickly, the U.S. Fire Service started questioning the logic of the ten a.m. policy and ironically implemented a new one that called for the kinds of "controlled burns" first used by Indigenous tribes. But the passage of time had dealt damage that could not be undone. Millions of acres of vegetation hadn't burned in decades, and by this time, much of this land was too populated to safely carry out controlled burns. And even if you could "rake up" all of California's dry vegetation, as President Trump once suggested, the state's radicalized weather cycles of intense rain followed by drought meant vegetation grew back, then dried out faster than humans could possibly keep up with. Man's manipulation of the land, stewarded by the tribes with such care, had run its course. Some of California's forests are unnaturally twice as dense today compared to 200 years ago, according to Frank's research. "We would call this an overdose," he told me. As researchers with the University of California put it, today's trees "compete with one another for sunlight and water and prevent smaller plants below from thriving. Stressed by drought and climate change, they are vulnerable to parasitic attacks. These dead trees are more fuel for wildfires, helping them spin out of control."

This "overdose" is where I want to drive home a point. Sometimes, someone will message me on social media about California's ancient history with wildfires—and the millions of acres that once burned—as proof climate change doesn't exist. "What's happening today happened hundreds of years ago," one person Tweeted, leaving out—or not knowing—other critical measurements. Wildfires in the early 1800s—while there were many individual ones—were mainly small, slow-moving affairs. Today's fires are fewer in number but far more explosive in scale. The top five largest wildfires in California's history have all happened since 2018 and burned hotter than ever before. According to Frank's research, fires in the 1800s killed about 10 percent of trees, with 90 percent surviving. The 2021 Dixie Fire literally reversed those numbers, and that was with the help of thousands of local, state, and federal firefighters using the best technology and tools to stop its spread. How many acres would really burn today if we chose not to lift a finger or—even worse—lost our grip?

For now, California is holding on tight, playing defense in an explosive and expensive game of whack-a-mole. The state's annual firefighting budget is more than $1 billion and growing. Recent numbers from NOAA, which monitors the amount of heat-trapping carbon dioxide in the atmosphere, don't give much hope for man regaining control. "The annual rate of increase in atmospheric carbon dioxide over the past 60 years is about 100 times faster than previous natural increases," NOAA wrote on their website. "Human activities have increased the concentration of carbon dioxide in our atmosphere, amplifying Earth's natural atmosphere." According to the U.S. Department of the Interior, these changes have seen the western fire season extend by at least eighty-four days in the last fifty years—and California's forests and the towering Wawona the Yurok, Karuk, and Hoopa helped grow are, for the first time in their history, feeling the heat.

There is something poetic about the name we've given to the largest of the sequoias remaining in the Golden State. At over 2,500 years old, the main attraction in Sequoia National Park enchants millions of visitors each year in a testament to the wonder and power of nature we've now put at critical risk in service to our short-term desires for convenience. I'd bet few of them realize "General Sherman," as it is now known, honors the man who brutally helped set into motion the series of events that now threaten its very existence.

When Giants Fall

"The last journalist we took to the top had a panic attack midway after telling us he was an experienced climber," said tree ecologist Wendy Baxter as I geared up to enter her laboratory, 250 feet in the air in the canopy of a giant sequoia in its namesake national park. Anthony Ambrose, her husband and research partner at U.C. Berkeley, had already made the ascent. The lean and rugged husband-and-wife team looked like they were ripped from the pages of a Patagonia catalogue. Anthony had a red beard and blue eyes. His head was covered by an orange Petzl helmet. Wendy's brown hair waved underneath hers. My visit was part of a report on the state's die-off of an estimated 129 million trees of various species, which scientists linked to the historic drought that had made them vulnerable to beetle infestation. Wendy and Anthony were here to figure out why the tree massacre hadn't yet hit the sequoias.

What neither knew, but perhaps suspected, was I had been a bit . . . *creative* with my climbing résumé. In fact, I've always had a wobbly-kneed, sweaty-palmed fear of heights, but said yes when my producer Chris Spinder asked me to do the story, rationalizing the reward of reaching the canopy was worth suffering through a day of crippling anxiety. In my defense, my (now long past) years as a Boy Scout had taken me on a few rock-climbing trips. Climbing a tree, I soon learned, was much different.

Before Anthony arrived at his treetop lab, he tied about 500 feet of fishing line to the end of a blunt-tipped arrow and, using a crossbow, launched it over the highest reachable branch. He then used that fishing wire to hoist up a sturdier nylon cord, which in return was used to pull up and over a heavier climbing rope. He anchored one end of the rope around a nearby tree and left the other end hanging free. "It's time to put on your harness," Wendy said as she handed me my gear.

Jumaring, in very basic terms, is a technique that involves climbing up a dangling rope, much like the ones in gym class, but with the assistance of autolocking carabiners, or breaks. These breaks attach the climber to the rope and prevent him from falling off or slipping down when his arms or legs get too tired and need a break. Anthony and Wendy looked like spiders gliding up silk as they made their way to the canopy. I, on the other hand, swung, crashed, and wobbled my way up, feeling more like an inchworm climbing an elephant's trunk.

It took me about an hour to reach Wendy and Anthony's eagle's nest in the sky, and the incredible panorama I found numbed the fiery adrenaline pulsing through every achy muscle. The forest appeared invincible from up above, but there was reason for concern. "These trees may be armored, but they do have an Achilles heel: their demand for water," Anthony explained from his cloud-hugging laboratory. "One sequoia uses up to a thousand gallons in a single day." Denser forests and drought meant trees were competing more for less water. But when faced with environmental stressors, Anthony explained, sequoias went into a form of hibernation. The behemoths could close pores in their leaves to conserve what little water they'd managed to collect through the worst of the drought.

What this conservation mode meant for the species' long-term health and that fireproof armor was not yet fully understood, but science did know that extended stress on a tree, even one as seemingly invincible as the sequoia, can make it vulnerable to disease and disarm its natural ability to keep bug infestation at bay. Hibernation also means sequoias wouldn't absorb carbon dioxide at the typical rate for conversion to food. Each year, the average sequoia removes more carbon dioxide from the atmosphere than 250 "normal" trees combined. "Our natural carbon filters are shutting off as the amount of CO_2 we release into the air increases," Wendy said before we repelled back down to the ground, gravity helping turn me into a spider. "It's going to take time and more research to understand the full impact on the trees and their habitat."

A partial picture did develop during the 2020 wildfire season. For the first time in recorded history, flames spring-boarded off an un-burned blanket of smaller invasive trees and reached a sequoia's fragile crown, once considered out of fire's reach, where they rapidly spread through the canopy. And because fires burn hotter than ever because of the more abundant fuel, even their two-foot-thick fireproof bark wasn't able to hold back the inferno. As many as 10,000 mature giant sequoias, representing about 10 percent of the species' population, were destroyed by a single fire that year, according to the National Park Service, which surveyed the aftermath following the so-called Castle Fire. The interior trunk of one was discovered still burning nine months later, and the following season, park rangers began wrapping the giants in flame-retardant aluminum fabric. One of America's most majestic sights now looked like a baked potato. If our ancient giants, a species that has outlived dinosaurs, are falling to their knees, what's the future for the relatively young towns and the hairless apes running around in their shadows?

The Uncharted Road Ahead

It's a dewy spring morning in 2021. Kylie doesn't own an alarm clock, and the volume on her old iPhone never rises above a mildly irritating whisper. Even if the phone did scream, it wouldn't matter. She usually tosses and turns herself awake at five a.m. anyway. That's when she quietly heads to the kitchen to make coffee. She doesn't need the caffeine, but the 20-foot shuffle across her small accommodations feels like a journey back to better times, and she needs it. Kylie pauses at the sink to look out the window at a landscape of inky silhouettes, remembering where the forest once stood. This predawn darkness is a canvas for memories painted with her eyes. Kylie loved watching the morning mist drift through her towering Sitka spruce and western hemlock.

The smell of hot water filtering through ground beans wakes Ellie, now ten years old, in what has become a bad habit neither can break. Now mother and daughter share not only their blue eyes and cherub cheeks, but for the last two and a half years the same bed and wake-up time. Every morning before sunrise they sit at the kitchen table, Kylie with her Folgers and Ellie with her cereal, and enjoy the one moment of the day that is pleasantly predictable. Here in their burrowed corner of California's northern woodlands, the passage of time is soothing.

It's the ticktock after this ritual, the world that awaits them on the other side of their front door, that keeps them awake at night. A new welcome sign with the same greeting, "May You Find Paradise to Be

All Its Name Implies," was planted where the old one once stood, but the new Paradise has a long way to go before it lives up to its name. "Sometimes I think that was all just a dream," Kylie told me. By now the sun's rays were catching floating dust particles by the kitchen window. The hour of "painting" was over. It was time to face reality.

"This is actually my parents' house," Kylie said as we walked out that front door and into the demolition zone that made up most of the neighborhood. Kylie and Ellie had moved out of their RV and into the home Kylie grew up in. The new living situation was far from ideal. The fire had damaged part of the three-bedroom house and police had placed a red sticker on the front door declaring it "uninhabitable," but Kylie had nowhere else to go and living in an RV was no longer sustainable. Kylie's father had boarded up the rooms where daylight broke through burned-out siding and singed insulation. "The Red Cross offered us a hotel room an hour away, but it was too far and I had no one to watch Ellie while I worked."

Kylie and Ellie were only supposed to stay for a few months while their house was rebuilt, but it had been more than two years and her contractor hadn't even broken ground. "They've given me the run-around. First, it was they were overwhelmed by demand. Then they said their major warehouse burned down in another wildfire." I spoke with her contractor, who confirmed the fire and said many of their own employees lost homes in Paradise and were working to help everyone rebuild as quickly as possible. "On dark days I think the holdup is a sign from God that it's time to leave," Kylie said.

"This is the road my school bus used to drive down when I was in high school," she pointed out as we came to a quiet country inter-section and began talking about her own crossroads. "People say now's the perfect time to get a fresh start, and I know they mean well, but it's not that simple. Because we've lost so much, we're holding on to what little we have, even if life looks like this. I never imagined we'd be wandering aimlessly the way we are. I promised Ellie a big slumber party once our house is built. It feels like a long way off, but I know it will be worth the sacrifice we're making now."

We walked around the neighborhood as I struggled to make sense of how little progress had been made toward repairing this community. Most of the half-acre plots where homes once stood had been cleared and covered in tall grass. Mother Nature was reclaiming what was hers. "Did you expect to see more?" Kylie asked, breaking the brief moment of silence.

"I did," I responded frankly, "but I guess I never thought about how much needed to be cleaned up first." It took more than a year to remove all the debris in Paradise. FEMA officials told me one thousand truckloads of material were removed every single day and taken to dump sites across the northern part of the state. Day by day, Paradise was slowly dismantled. It made me think of an old-fashioned Polaroid working in reverse, the colors returning to the pouch in the base of the instant film, erasing the image they had once revealed.

Rolling the Dice

Paradise had taken 170 years to build before it burned away. In 1841, at the age of twenty-two, John Bidwell left his family's home in Erie, Pennsylvania, and joined a wagon train embarking on a vanguard journey first carved decades earlier by trailblazing explorers Meriwether Lewis and William Clark and made iconic by the generations of pioneers called west by promises of opportunity.

To call the trip risky is an understatement. Travelers withered in Nebraska's sun-bleached savannahs and froze in Idaho's icy passes. Bridges were often rotted and many rivers had none at all, meaning wagons had to be guided over rocky riverbeds. Then there was an alphabet of diseases, from cholera and dysentery to measles and smallpox. Nearly one out of every ten travelers died, according to the National Park Service. Government marketing downplayed the risk, but the roughly six-month trek was like an expedition to another planet.

Bidwell was among the estimated 400,000 people who tried. His expedition splintered off from Lewis and Clark's trail near Idaho's Fort

Hall where they forged the first major overland trek through Nevada's 40-mile desert and rugged ranges to reach California. Bidwell's journey along the Feather River helped usher in the modern-day gold rush that transformed small trading posts into thriving ecosystems. According to the city's website, Paradise may have gotten its name from a local saloon where pioneers drank and gambled. Its name was Pair O' Dice.

In the first two and a half years after the Camp Fire tore through Paradise, only an estimated 4,700 of the town's 29,500 residents returned, most of them to homes that didn't burn down. In the same stretch of time, the town hall only received 1,618 building permit applications for destroyed properties, according to the mayor's office. Despite quickly approving most of them, only 753 homes had been rebuilt. Town officials were trying to track down other residents to see if they planned on coming back. If people weren't returning to Paradise, the lost property tax revenue would paralyze rebuilding efforts.

"We've got enough federal funding to support the city for the next twenty years," Mayor Steve Crowder told me in front of his newly built home. He was confident Paradise would return stronger than ever before, even though the next numbers he threw at me didn't add up. "We estimate 50 percent of the town will return in fifteen years," he said, recognizing the surprise in my eyes. "Progress is slow. It's been nearly three years and we're still removing dead trees. Returning isn't for the faint of heart, but Paradise is full of fighters." Like so many of those who returned to Paradise quickly, Crowder was born and raised here and couldn't envision living anywhere else. When the firestorm swept in, he chose to help police guide traffic out, even while receiving reports from the fire department that his own home was engulfed by flames. His effort to now convince people to return was tireless. Not everyone was willing to roll the dice.

To Stay or to Go

Kitchens are rooms with a future.

To peek inside someone's refrigerator is the closest I've ever gotten to mind reading. Every item is a sign that, no matter how glass-half-empty a person might be, we are all filled with some measure of optimism that tomorrow will come, and when it does, we will be nourished.

Heavy cream. Strawberries. Condensed milk. Sliced turkey. Swiss cheese. Pork. Applesauce. Chilled white wine. A fresh gallon of milk. Katy Schrum's kitchen had all the ingredients of hope. It also hid the kind of past that makes her uniquely aware of how tomorrows aren't guaranteed. In a cupboard near her oven, you'll find a different list taped to the back of the door:

Norm. Sami. Molly. Ezra. Logan. Jenny. Susie. Lisa. Camellia. Brian. Bryan. Kristi. Jesus. Pam. Suzanne. Pastor. Great Grandpa. Grandma Taylor. Grandma Brown. Mom. Dad. Amy. Rick. Leo. Paul. Jimmy. Lori. Aaron. Suzanne. Linda. Aunt Hazel. Rex. Peter. Lynn. Melissa. Michelle. Marie. Glen. David. Mary. Gary. Emi. Margaret. Willy. Mom. Dad. Joyce. Alan. Robbie. Barb. Cynthia. Colton. Anakin. Korbs. Aunt Jackie. Uncle Gene. Great Grandma Green. Laura. Taylor. Shauna. Jim.

They are the names of her friends and family from Paradise. Some have passed on, but most are alive. "I look at that list, pick a name, and spend the entire day thinking of that person. It's how I remember and hold on," she told me with a somber pause. While she no longer lived in Paradise, her mind swung in a birdcage gilded with nostalgia. "I've gone weeks without pulling a name off the list." It was getting harder for Katy to revisit the world she was forced to leave behind.

I first met Katy two years earlier on the same trip where I interviewed Ellie and Kylie. Katy was Ellie's teacher, conducting class for second-grade students in a borrowed classroom in a neighboring town. Paradise Elementary, all of it, had been destroyed in the blaze. Even the file cabinets were melted into metallic puddles. "Only one-third of the students have returned," she told me back then. Many moved with family out of town or even out of state. At the time, Katy wasn't sure what she would do. She and her husband had lost their home and were renting an apartment near town. Katy was the kind of teacher that children ran to with excitement each morning. Their love for her rivaled the love they held for their own parents. She had been at Paradise Elementary for twenty years and had watched some of her students go on to have children of their own. Thirty-four of her thirty-nine students in 2018 lost their homes, and trauma permeated every moment of life in Paradise after the firestorm, including her class's lessons. Every morning after the fire, as each remaining student entered the classroom, Katy asked them to choose either a red- or green-colored Popsicle stick. Red for sad, green for happy. The students who picked red would later be treated to cookies and a chat with the school therapist. "They are just like the adults. One day they are doing fine, and then out of the blue something triggers them and they need help. Last week one of the boys looked at me and said 'I want to go home' and I told him 'Me too but we couldn't.' It broke my heart."

The week of my visit with Katy, the results were split fifty/fifty between red and green. "Some days every student picks red. Today is a great day," she said. But great days were fleeting, for the kids and for

Katy. Eventually the trauma of driving through what little was left of Paradise wore her down. She was losing hope. Finding it again meant leaving.

"I feel like I can breathe again," she now told me on the phone from her new home in Eureka, California—the one with her list of names taped to the back of her cupboard. "It's been going well, but then something will remind me of Paradise and I'll struggle to snap out of it. Just hearing your voice put a lump in my throat and tears in my eyes," she told me. I apologized for disrupting her day and told her I could call her back, but she said it was good to be in touch. "It's so important to remember, but it can also be really hard. This is a good reminder." For Katy, the months leading up to the Camp Fire had seemed too good to be true in that "pinch me, I can't believe how lucky I am" kind of way. Her twenty-six-year-old daughter had just moved into a home in Magalia, the next town over, and had gotten a job as a fifth-grade teacher at Paradise Elementary. They were working together, under the same roof. Katy's husband had retired from his job early to stay home and look after the grandkids. It was the kind of tightly woven family tree that hand-sewn heirloom quilts are made of, until the fire burned every fiber.

Katy stayed in Paradise for as long as she could, until the pain was too much to bear. She was tired of picking the red Popsicle stick. "Home should be a place to live without fear and without pity and without judgment." It reminded me of something a psychologist and minister named Dr. Thema Bryant once wrote that has stuck with me for years:

"May you find and create spaces where you don't have to fight, defend, anticipate, perform, prove your humanity. May you have space to breathe, be, create, and bloom. Let it be so."

Finding that space isn't easy. Katy often wonders what life would have looked like had she stayed. "I can never predict what's going to trigger me. The other day I put on my shoes to go for a jog and started crying. I curled up into a ball and had a good cry." Katy loved Paradise, but knew it wasn't good for her. "I just don't know how people return to it. No matter how much better you rebuild a town, can anything hold back another wildfire?" she asked. "I think today's a good day to pick another name off my list," she said before we wished each other a good day and hung up.

Yes, cleanup would take time, but I was beginning to see Paradise's struggle to rebuild in a more sobering light. To move back to Paradise again would mean being anxious six months out of the year and probably not sleeping if the wind blew. Maybe there simply weren't enough people willing to reinvest both financially and emotionally, and leaders, business owners, and residents were pulling back in a form of self-preservation. It also didn't help that funds to rebuild were slow to come.

On June 16, 2020, PG&E took responsibility for the Camp Fire. "Our equipment started that fire," said PG&E CEO Bill Johnson at a dramatic hearing attended by victims' families. Johnson admitted that faulty electrical transmission lines sparked the blaze that was then carried by the wind. "PG&E will never forget the Camp Fire and all that it took away from the region," he said on the stand in one of his last public appearances before retiring. In a settlement, the company agreed to pay $13.5 billion to residents and business owners impacted by the fire and other wildfires from the previous year linked to the company's failed equipment. "Compensating these victims fairly and quickly has been our primary goal throughout these proceedings, and I'm glad to say that today we funded the Fire Victim Trust for their benefit," interim PG&E CEO Bill Smith said fifteen days later. But an investigation by local public radio station KQED found that, one year later, most of the funds in that trust had yet to be distributed because of bureaucratic red tape, according to public records they reviewed. "We found that in its first year, the trust had racked up $51 million in overhead, and distributed only $7 million to fire victims. Ninety percent of outgoing funds paid lawyers and consultants in 2020 while the vast majority of fire victims waited for help." That included Katy and her husband. In all, "fewer than 3% of fire victims—1,867 out of approximately 70,000 total—have had their claims fully processed," the report said. Debt, along with memories, haunted many.

For those who didn't lose their homes in the Camp Fire, life in Paradise could still be confusing and painful. Resident Jessica Distefano stood outside her house with me—the dream home she and her husband built was one of the only ones to survive in her neighborhood. "At times I wish that it was gone," she told me. "When it's windy the ash blows and we don't go outside because who knows what's in the air. We can't drink from the tap or bath because the fire melted all these metals and plastics that released toxins into the water." Every month, a water truck filled a massive plastic tank that was parked on the side of her driveway. She was not only concerned about her family's health, but also their long-term safety from another megafire. For her, like for Katy, it wasn't about if, but when another inferno would reach Paradise. "It's only a matter of time," she said.

For Jessica, it was a catch-22; stay and possibly face off with another fire down the road, or sell at a loss. She had considered putting her home on the market, but real-estate deals following the Camp Fire showed some properties selling for 50 percent less than the price set before the fire, and property values in nearby communities undamaged by the flames dropped 15 percent. One developer described it as a "fire sale" to the *Sacramento Bee*. Jessica was part of a growing group of people asking FEMA to buy back land from owners. For decades, the federal agency, using taxpayer dollars, has been purchasing land from flood victims like in Staten Island after Hurricane Sandy swept through and drowned coastal homes, but no such plan is in place for wildfires, which only recently began wiping out entire communities. FEMA has only committed to studying whether such a plan would be financially feasible. There's also the question of where all the unhoused would go in a state with finite housing options, and what the emotional, social, and cultural toll would be for these climate refugees and their host communities. Holistically—and also practically—is it better for victims of extreme weather to surrender or stay?

"We would never tell someone what they should do. That's a deeply personal question. But overall I find most people want to return home and there's a small group who are so traumatized that they want to leave," Jennifer Gray Thompson told me. She founded the nonprofit organization After the Fire to help victims of wildfires rebuild or relocate. "There's so many variables involved with staying, but I've found the yearning to return is almost universal, even if it could take five or ten years to return. For those that have grown up in a place impacted by a disaster like a fire, home is more than just a house. It's about the community, where you had your first kiss, your first ice cream cone. That sort of thing doesn't have a value," she said.

A 2021 Marist Poll found two-thirds of Americans would rather rebuild than relocate if their home was destroyed in an extreme weather event. "Our goal is to help these people make the right decisions. If they chose to stay and rebuild, how can they do so in a way that is smart? In a way that can reduce fire risk in the future?" Jennifer said. "Fire is pervasive in the West. I know one woman who moved from Paradise after the Camp Fire only to be threatened by another fire in Oregon. And here's the irony: In many cases the safest place to live after a fire is the place that just burned down. There's nothing left to burn. And because these towns are starting from scratch, there's an opportunity to fireproof them in ways you couldn't before. At some point every community is going to have to learn to live and adapt to climate change and the threats."

She had a point. Remember, one in ten homes across America are feeling the impacts of climate change, and in California, 41 percent of properties are considered at risk of being impacted by a wildfire, according to the research group First Street Foundation. If you're going to go head-to-head with Mother Nature, the thinking is, you might as well have the home-field advantage. Later we'll meet one community with such an advantage that has survived storms of a different nature, as designed. But more and more, in what will become a recurring theme in this book, people are rolling the dice and willingly moving into land in defiance of Mother Nature.

As Paradise struggles with their own adaptation, and residents like Jessica and Katy struggle with how best to rebuild their lives, wildfire-threatened communities across the nation's West are actually exploding in popularity and most have not been retrofitted to withstand today's megafires. As of 2020, more than 16 million homes were located in fire-prone areas near forests, grasslands, and shrublands, according to joint research published by the University of Wisconsin-Madison and the U.S. Forest Service. That's a more than 60 percent jump from 1990. The wildland-urban interface has expanded for a variety of reasons, including more affordable land and people wanting to be closer to nature, and no state has experienced more rapid growth than California, where more than five million homes have been built on wildfire land in that time span. Western states could impose tighter restrictions on future development, but experts say that's unlikely because so much of the West is at risk of catching fire, and a growing population requires space. Meanwhile the proliferation of new homes continues to make it more difficult for firefighters to protect life and property. Officials not only face the real possibility of a "controlled fire" burning out of control, but also expensive legal battles from homeowners who are increasingly saying "not in my backyard" because of those risks.

Living in America's West has become a game of odds. Ultimately, for better or for worse, the insurance industry may dictate who lives where. In 2023, State Farm joined Allstate, American International Group, and Chubb to announce they would stop issuing new policies covering homes in California, specifically citing the risk of wildfires. And for the insurers that are sticking around, some rates for homeowners more than doubled practically overnight. "I call it gentrification by wildfire," said Johnson. "Those with money will find a way and they can fortify their homes, but what does a community lose—how do the identities of these towns change—when most can't afford to stay, or in the aftermath of a wildfire chose not to return?" Climate change is transforming the lives we have taken for granted into luxuries many will be forced to live without if action isn't taken now.

"Nature's Way"

In 2008, California adopted some of the strictest codes in the country for new homes in high-risk fire areas, including fire-resistant materials, larger defensible space around homes (this involves clearing brush and trees), and easy access to water for firefighters. Notably, homes built before 2008 are not required to be retrofitted. Even so, a *Los Angeles Times* analysis of assessor records and fire surveys showed homes built to code had just a 13 percent survival rate in the Camp Fire. If communities already hit by wildfires are some of the safest to live in, as Jennifer Gray Thompson put it, there's clearly a lot of room to build back smarter. Newer technologies, including fire-resistant paints and rooftop sprinkler systems, have become mainstream in recent years and offer additional protection at a lower cost. But what about measures that don't require a homeowner to open their wallet in the first place? One of the most effective solutions in stopping fires from even breaching communities may come from Nature herself.

We are a uniquely destructive species, and the only one on the planet capable of pushing another one to extinction. From habitat change (including overfishing and overhunting, poaching and pollution, environmental destruction and engineering) to climate change, there are now up to one million species at risk of vanishing thanks in large part to our behavior, according to a 2019 report by the Intergovernmental

Science-Policy Platform on Biodiversity and Ecosystem Services, or IPBES. The North American beaver was once on that list.

As many as 400 million beavers once puddled North America, according to the U.S. Fish and Wildlife Service, doing what God put them on this Earth to do: chop down trees and build dams. The water those dams pool helps create vibrant wetlands packed with the kind of biodiversity that keeps ecosystems resilient. It's why beavers are called a "keystone species," meaning they help glue the entire ecosystem together. But early settlers found different value in America's largest rodent. Its thick pelt is among some of the softest in the animal kingdom and naturally waterproof, which made it perfect for jackets and hats in a rapidly developing world where consumers increasingly demanded style along with substance. During the first several centuries of the European colonization of North America, beaver pelts were one of the New World's most lucrative exports, especially after European nations hunted their own native beaver populations to near extinction. A similar pattern of gluttony soon depleted the eastern American population, and it can be argued that man's addiction to beaver fur was actually responsible for sparking America's western expansion. The little-talked-about California Fur Rush predated the golden one by more than twenty years.

By 1900, the beaver population had been whittled down to an estimated 100,000. And the extermination didn't end there. Trappers, by invading the mammals' often remote habitats, paved the way for settlers to develop these once off-limits slices of land. This came in addition to beavers killed as a "nuisance," with many clearing the species as a preventative measure against flooding caused by their dams. The U.S. government eventually approved a series of federal protections as potential extinction became a real possibility, and over time the population rebounded, ever so slightly, as beaver fur went out of fashion. Today roughly 10 to 15 million beavers roam North America. While that's a fraction of the original population, it's considered such a conservation comeback that landowners are once again allowed to trap and even kill the nuisance critters. But a recent discovery is now placing greater value on the North American beaver alive, instead of dead.

Dr. Emily Fairfax, who studies beavers in the field and refers to them as "ecosystem engineers," couldn't believe her eyes when she went to check on a beaver pond following a wildfire in 2020. The beavers' dam, and the lush oasis it created, were able to hold back the state's radicalized fire and survive the blaze. "I was like, *this is unreal.* Does this happen everywhere?" Emily said as we slipped into waders and boots so she could take me to see what she called a beaver masterpiece in Atascadero, California. Emily had long blond hair tied back in a ponytail, and an orange-slice smile that widened once knee-deep in water.

"This is a beaver dam," she pointed out as we slushed through a shallow stream to a 10-foot-tall webbed wall of sticks and mud that held back a large pond. "A family of beavers built this dam, which over time has collected an absolutely enormous volume of water. Every day a little is released into the stream, creating this unusual strip of green. I mean, we're in a microclimate here. It must be ten degrees cooler than anywhere else in the area. This would all be bone dry and sandy, which is what it was before, if the beavers hadn't moved in." It was the only greenery for miles.

"How are beaver dams better than, say, hydroelectric dams that provide clean energy?" I wondered.

"That energy comes at a cost," Emily said. "Hydro dams are walled-off concrete fortresses that can cut off the entire flow of a river, leading to droughts downstream and disrupting species migration and freshwater biodiversity. I think of beaver dams as speed bumps," she said. "Beaver dams are nature's way."

As Emily explained, when a beaver builds a dam, the structure slows rainwater and snowmelt from rapidly draining down rivers into oceans. The result is a natural reservoir capable of storing water for years while still supporting wildlife migration. "When the beavers move in here and they slow this water down, a lot of it goes into recharging the groundwater, and that's what we're pumping for irrigation. That's what we use for our food, that's what we use for our lawns. And these beavers are recharging it for us. So they're sort of depositing water into the bank that we take out at a later date." The drier an area is, the more critical dams are because soil, over extended periods of time without water, becomes too brittle to retain water when it eventually arrives. It's like watering parched soil in a potted plant. The water simply flows right out of that hole at the bottom, not giving the roots time to hydrate. But if you obstruct that hole, the water sticks around, replenishing the soil and the roots. The results in the natural world are wetlands, which are breeding grounds for about 80 percent of animal species in the American West. According to the Fish and Wildlife Service, California's wetlands have shrunk by 90 percent since the 1800s, around the time beavers started vanishing from the landscape. "The desertification of California aligns with the loss of our beaver population. The science isn't clear, but I think it's safe to assume wildfires wouldn't be as bad as they are today if we had more beavers," Emily said. "And if we can keep vegetation alive and prevent fires, we can also prevent other related environmental disasters like mudslides and extreme flooding."

When a wildfire is extinguished, it leaves behind a burn scar—dandruff layers of charred soil up to five inches thick that, like the dried soil in a potted plant, are terrible at absorbing water and incredibly vulnerable to erosion because there are no living roots holding the earth together. Now add in the impact California's warming weather is having on the behavior of the state's winter wet season that immediately follows fire season. Today, more rain is falling instead of alpine snow, which melts slowly over time. It's a deadly mix of ingredients. A deluge on a burn scar in Montecito triggered a mudslide on January 9, 2018, which killed twenty-three people. Many of them were buried alive. And in the first few months of 2023, a series of atmospheric rivers—rivers of moisture in the sky carried by air currents known as jet streams—dumped 32 trillion gallons of water and placed 90 percent of the state under flood watches, according to the National Weather Service. In Northern California, a month's worth of rain poured on a burn scar in a single day. This violent runoff was so powerful, it caused mudslides and landslides, quickly inundated rivers, and shattered a levee near Galt, California, that had recently been restored to last until 2050. In a matter of minutes, entire communities were swallowed by floodwater.

I'll never forget the man who came up to me after I reported for Norah O'Donnell on the levee break. My team and I went live from New Hope Road, which had been flooded along with the vineyard it ran through. "By any chance have you seen a gray sedan on the side of the road while driving around today? My sister-in-law disappeared during the storm and we haven't heard from her," he said. He and his family had spent three days driving down country roads in a desperate search. "I'm sorry, I haven't, and this is as far as we were able to go down this road," I explained before his eye caught the top of what looked like a submerged car a hundred yards away, partially concealed by vines. My team and I had no idea a vehicle was even there. From our location, we watched as the man waded into the water and out to the car. He was too far away to hear, but I could read his body language as he slumped over, the palms of his raisined hands on his shaking knees. His grief rippled the stagnant floodwater around him. Katherine Martinez was sixty-one years old when she drove off the road during the storm and drowned in its surge.

Water in a state that needs more of it should be a blessing, but not when that same state is so dry it can't store or even handle what's fallen. At the time of my reporting from the flood at New Hope Road, the state had been planning a new series of water storage projects, including expanding reservoirs, but saw the adaptive initiative held up by permitting issues for nearly a decade. In Los Angeles County, voters—recognizing the state's historic drought—approved extraordinary steps to avert crisis years ago. The Safe Clean Water Program, or Measure W, was passed in 2018 and allocates $300 million annually to projects that include adding cisterns and more green space to trap and absorb rainfall, while also removing asphalt and other "hardscape" that acts like a water park slide by shooting runoff into the Pacific instead of allowing the ground to absorb it. But like many things that involve government hands, Measure W has failed to make a splash. An investigation by the *Los Angeles Times* found that in three years, the county had only set aside thirty acres of land for conservation efforts. Californians—like a growing number of Americans—want action, but their leaders are hanging them out to dry. And the holdup has proven costly in more ways than one. The nearly 36 trillion gallons of water that fell in the winter of 2023—that's more water than what flows through the Amazon River every day—wasn't enough to bring a permanent end to California's drought. Much of the water was rapidly flushed into the Pacific.

While the West's historic drought, wildfires, and mudslides have led to a reckoning and wanting of beavers, with so many of them removed from the landscape, reintroducing them now requires help from humans. And that's how I first met researcher Nick Bouwes who, upon speaking with my producer Christian Duran, asked if we'd be up for building a beaver dam for a ranch owner in Utah and then releasing a few into their new home. Who says no to that?

Nick is the head of the Beaver Ecology and Relocation Center based out of Salt Lake City, Utah, and oversees a team of volunteers who travel across the West building "starter dams," what's technically known in the emerging industry as beaver dam analogues. That's how I found myself waist deep in a babbling creek on a mountainous ranch in Coalville, Utah, slinging handfuls of mud over layers of tree branches and sticks. The landowner had heard about Emily Fairfax's research and Nick's relocation project and wanted in. "Too much water can be a bad thing, but when there's none of it and crops are dying and ranchers are selling off their cattle and there's no end in sight, well, that's enough to change people's opinions," Nick said as he grabbed a pile of mud from the bank of the creek. "Beavers, what they do is they get in here and they scoop the mud up, they just come and grab a whole bunch, push it with their chest and hand and it into the crevasses. I would get on my belly and push it in, but I don't want to do that yet," he chuckled.

Nick had a thick crop of brown hair and a stubbled face. His brown eyes, which squinted even in the shade, remained focused on our project. He looked all business, but the occasional crack of a smile let me know he didn't take himself too seriously. It amazed me how much work went into something that should be—as it had once been—natural.

"A lot of these streams, if it's too shallow, beavers can't defend themselves. They're just kind of like walking hot dogs for predators. They need deep water to swim," Nick said as he handed me more branches that had been freshly cut by his volunteers. Decades without beavers meant there weren't deep enough streams to relocate them to without human intervention.

Building a beaver dam is wet and dirty work, but fortunately not an exact science. It took about thirty minutes to create a pool about four feet deep. "This should be a safe space for the guys we're releasing today." Those "guys," a male and a female if I'm being accurate, were resting in cages kept in the back of an air-conditioned van. Beavers are nocturnal animals. It's rare to see one, let alone see one just a few feet away. I was excited to get an up-close look.

I stared into the black, pebble-sized eyes of the beaver we'd soon be releasing. He looked like a cross between a guinea pig and a sea otter and was surprisingly cute. I've always loved animals, yet never gave the beaver much thought before. Why? "He's about ninety pounds, but at least he's not kicking around," Nick said as he wired the cage the beaver was in onto the wooden arm of a shovel. I grabbed one end and Nick the other and, in what looked like a makeshift gondola, we airlifted the beaver down the steep canyon to the analogue dam we had just built. "This guy came to us from another ranch, where it was causing trouble."

I was still looking into our beaver's tiny blinking eyes and could tell he was confused. Instinctively he must have known most beavers, once in cages, don't make it out alive. If we are to believe the sounds and gestures of an animal allow them to communicate on a basic, primal level, then it seems logical to think this guy's mother warned him of humans, just like his mother's mother warned her. That probably explains why, when we opened the cage near our handmade dam, he looked at us for a second before cautiously waddling out and slipping into the stream. The water flowed over his thick coat like melted chocolate. A pair of volunteers released the female beaver a few yards upstream. "The goal is for them to meet naturally and colonize this dam together. The odds are good for them here," Nick said as we watched them inspect their new home, and later, each other. "They look happy." More than one thousand beaver dam analogues have been successfully resettled in the West, with hundreds of more requests coming in from farmers and ranchers. "At some point in history people just accepted beavers were commodities and pests and never stopped to ask what removing them would mean for the environment and our own lives. We trapped and killed them out of some kind of necessity and now we're working twice as hard to reintroduce them because we've run out of all other ways of restoring our planet," Nick said.

While scientists say reducing our greenhouse gasses will help stabilize our climate, they warn that, even if we were to cut all emissions right this second, it would still take decades if not longer to see positive results in our climate. That's because the impacts of our pollution operate in a delayed feedback loop. It takes time for us to feel the effect of the greenhouse gasses we've pumped into our ecosystems. And the same is also true when trying to lower this fever. Climate is delicate and operates at a slower pace than habitat, which is much more resilient.

As we invest in the future by cutting back greenhouse gasses, we can also invest in strategies that have near immediate impact. As I watched our two beavers bob in the water of their new home like they had lived there all their lives, it was clearer than ever that restoring the land and learning to coexist with nature could be the fastest, most cost-effective, and lasting way to make our communities—our ecosystems—resilient to failure.

A Home Called Faith

In the fall of 2022, following our report, the California Department of Fish and Wildlife posted its first job listing for the state's new beaver restoration program, inspired by Emily and Nick's work. California's new wildfire mitigation strategy pumped millions of taxpayer dollars into beaver relocation efforts. "There's been a major paradigm shift throughout the West where people have really transitioned from viewing beavers strictly as a nuisance species and recognizing them for the ecological benefits that they have," said Valerie Cook, the program's director. The additional state funding will go a long way to helping reintroduce beavers to their natural habitat, but let's just hope politicians don't find a way to drain resources and hold up what I saw as a streamlined process.

There have been discussions about releasing beavers into Feather Creek near Paradise, California, as part of a multistep effort to rebuild the ecosystem and ward off the next megafire. After all, a moat of habitat-nourishing water surrounded by healthy vegetation does seem like a pretty effective fire line. But discussions haven't gone far. For now the focus remains on trying to return people to the land. And those who have come back seem to believe human engineering is the best path forward. In recent years, a new kind of home has started to replace some of the wooden cabins that were once nestled under the tree canopies. "Q cabins" get their name from the Quonset huts that were manufactured during World War II to hold military supplies. They're half-circle structures made of noncombustible steel and sort of look like an airplane hangar. According to their manufacturer, they can withstand heat up to 2,600 degrees Fahrenheit, though fortunately none have had to be put to the test in Paradise.

A handful of these homes have risen in the town, and they are beautiful. There's a sleek, modern simplicity to them. But with few trees still standing after the Camp Fire, and with many of the surviving trees chopped down to make defensible space, there's an eerie emptiness to the town. In some areas, the Q cabins are the only things rising from the land. What used to be a forest now almost looks like a desert. Paradise officials have also updated their faulty CodeRed system, adding sirens to the town, similar to the ones used during the Cold War when America faced a nuclear threat. To me, and I know this is going to sound critical, this new landscape almost looks like humans are at war with Nature, instead of working with her to fight off ecosystem collapse.

Our friend Kylie's new home, a simple one-story wooden ranch house that was built near the footprint of her old house, does not conform to the trend in town. She couldn't afford one of the Q cabins. Paradise—its second act at least—is beginning to show signs of "gentrification by wildfire." It's an inequitable new world divided into the haves and have-nots; the rich and the poor; the white, black, and brown. And while this world is better adapted to fight off Mother Nature in the short term, it does not coexist with her. Man's grip will soon slip if history is any indication. It doesn't have to be this way. Californians have asked for change . . . they've even voted to fork over more of their own salaries to fund preemptive measures. So far, they've been let down. As I've crisscrossed America's wildfire west, I've seen firsthand how it *is* possible to protect ourselves from disaster, even in the most threatened spaces. But the first step requires immediate action so our communities can not only survive but remain the diversified ecosystems we call home.

For now, Kylie takes comfort in the more modest adaptations she was able to afford, like clearing her property of all its trees and dry vegetation. "I do think things will get better now that we know how bad things could get if we don't work at it. There's not a lot left to burn," she said. It took more than two years before Kylie and Ellie could move into their new home. They were handed the keys on December 9, 2021, and were looking forward to getting a Christmas tree and decorating it with new ornaments and finally having that slumber party Kylie promised Ellie. Before walking through the front door for the first time, the two looked at each other and saw the future flash before their eyes. They took a moment to whisper a prayer and came up with a name for their new home. From that day forward, she would be called "Faith."

PART TWO

WATER

Ice Chaser

You'll only find a handful of photos of John Mercer online. One of them, from his 1987 obituary, shows his mischievous gap-toothed smile and warm almond-shaped eyes framed by thick black glasses that disappear under a chunky cable-knit cap. Another one, taken in front of a snowy campsite, captures a much younger, though bleary-eyed, Mercer on the frigid front lines of groundbreaking research. John couldn't have known then, as he sat hunched-over in front of his tent drawing warmth from a tin cup of coffee, that the data he was collecting from the most remote corner of the world—an endeavor that earned him ridicule as he collected it in life—would be looked to in fear long after his death.

"If the apparent warming trend is real and continues . . . whether because of industrial pollution of the atmosphere or for any other reason, the unstable West Antarctic ice sheet will become a threat to coastal areas of the world," Mercer concluded in his now-seminal 1968 paper on the state of our planet's ice. But back then, few took note. You could blame it on the less-than-eye-catching title "Antarctic Ice and Sangamon Sea Level," or maybe just bad timing. At the time the eyes of science were focused up, on the sky; up, enraptured by the *Apollo 11* mission, America's then-unfathomably big swing at declaring victory in the Space Race by placing a man on the moon. But with eyes tipped toward the ground, John trained his focus here, on an environ most alien in its own way. Our planet's north and south poles and the role they so quietly played in stabilizing our biosphere were more mystery than proven science back then, and "climate change" was a largely untested concept whispered about in small rooms and the far reaches of the academy.

John's love for ice began as a student at Cambridge University in England where he was born and raised. He studied geography under Professor W. Vaughan Lewis, who was one of the original members of the International Glaciological Society, back when the club was a nation of one: Britain. Vaughan believed glaciers, from the Himalayas and the Rockies to the North and South Poles, were apex habitats responsible for circulating incredible volumes of nutrients to other ecosystems, much like a human heart pumping blood through the body. "The grandeur of the mountains, of the ice they bear and the streams they nourish, is for us all to study," he once wrote. But Vaughan was born ahead of his time, when traveling to our planet's most important bodies of ice was difficult if not impossible. John was determined to follow in his professor's academic footsteps before paving his own wild path.

He continued his studies at McGill University in Montreal and eventually settled at the Institute of Polar Studies at Ohio State University, where he remained until his death. For more than two decades, he struggled to secure funding for frozen expeditions that lasted months in some of the planet's harshest conditions. He led teams to northern Canada's Baffin Island and Greenland, but focused most of his attention on Antarctica, where that grainy campsite photo was taken. He was known by his colleagues for his tireless commitment to his work and his eccentric work habits while out on the ice. It was rumored John preferred to gather field measurements in the nude.

John's 1968 paper became the foundation on which he built the rest of his career's research. Back then he determined that the Antarctic ice sheet, considered unmeltable by most in his field, had thawed out 120,000 years ago during an interglacial period known for its climate instability. Sea levels rose, he believed, by nearly 16 feet as a result. It was, by itself, a major revelation, but nothing compared to the peer-reviewed paper he published a decade later in the science journal

Nature. And with the title "West Antarctic Ice Sheet and CO_2 Greenhouse Effect: A Threat of Disaster," many began to look at what was unfolding down in the Antarctic. John's research showed that human activity was driving CO_2 levels to rates not seen in millions of years, so fast there was no time for Mother Nature or humanity to adapt to the unintended consequences. By comparing newly available satellite imagery from NASA with his land surveys, he further concluded the West Antarctic Ice Sheet had already begun to melt like it had before. Earth was in the early stages of crisis.

"If present trends in fossil fuel consumption continue, and if the greenhouse warming effect of the resultant increasing atmospheric carbon dioxide is as great as the most advanced current models suggest, a critical level of warmth will have been passed in high southern latitudes 50 years from now, and deglaciation (reduction) of West Antarctica will be imminent or in progress. Deglaciation would probably be rapid once it had started, and when complete would have led to a rise in sea level of about five meters [16 feet] along most coasts."

His warnings were now an alarming call to action. Humans had fifty years to rein in the levels of greenhouse gasses admitted into the atmosphere, or temperatures would spike, ice sheets would collapse, sea levels would rise, and anyone living on a coast would be—in a word—screwed.

"This deglaciation may be part of the price that must be paid in order to buy enough time for industrial civilization to make the changeover from fossil fuels to other sources of energy," he wrote, offering a measure of hope. But his words went unheeded.

Other glaciologists examined John's models and determined the data backed up his findings, but to his surprise and ultimate downfall, far more publicly dismissed the idea that an ice sheet three miles deep and twice the size of Australia could significantly disintegrate so rapidly. John was assailed for being an alarmist and, overnight, funding for his work froze.

James Hansen, renowned longtime director of NASA's Goddard Institute for Space Studies, would call it the "John Mercer effect." As he saw it, scientists were chastened by John Mercer's downfall and began sugarcoating the results of their climate research to avoid scaring away the funding needed to carry it out in the first place. "It seems to me that scientists downplaying the dangers of climate change fare better when it comes to getting funding," James wrote in an alarming 2007 article for *New Scientist*. "Drawing attention to the dangers of global warming may or may not have helped increase funding for the relevant scientific areas, but it surely did not help individuals like Mercer who stuck their heads out."

But the ultimate loser, he wrote, would be the public. "We may rue reticence if it means no action is taken until it is too late to prevent future disasters." In 2014, thirty-six years after John Mercer first sounded his alarm, two separate field teams found the West Antarctic Ice Sheet was not only melting, but irreversibly so. And then, in 2021, satellite images revealed severe cracks in the Thwaites Glacier, an Antarctic ice shelf the size of Florida. Reports from that same year measured heat-trapping carbon dioxide molecules in the atmosphere at 412 parts per million, their highest level in 3.6 million years. It was the atmospheric equivalent of a failed Breathalyzer test. Scientists warned that in this drunken state, the Thwaites could shatter completely by 2026. And sea level rise wasn't the only concern. Perhaps worse, the influx of freshwater from the thawed ice sheet would throw off the chemistry of the world's oceans, altering currents, impacting wildlife, and radicalizing storms (we'll explore this later). John Mercer's wild, career-crippling fifty-year prediction was in fact conservative, by two years. But even *he* couldn't fully envision the domino effect a crack in the Antarctic ice would have on our planet's oceans, and the megastorms they would birth.

The Wake-Up Call

The alarm on my cell phone went off at 2:30 a.m. For a second I was confused about where I was and why I was lying on a couch. The power had gone out, but the rain-refracted light from some unknown source outside poured through the windows in a splotchy sepia tone too weak to fully pierce the darkness of my foreign accommodations. The wind strummed the siding of the house in a rhythm I've learned, in the years to follow, to find soothing—the final eerie chords of a passing storm. The phone's angry alarm snapped me into reality, which included the realization that I'd forgotten to charge it before the power was knocked offline. Thirty percent battery would only last me an hour at the rate I knew I'd soon be working.

I felt around in the dark for my backpack. It's part safety blanket, part insurance plan. Inside I have every unremarkable tool I need to function in the field. Nuts, beef jerky, granola bars, and a can of sugar-free Red Bull for when my energy starts to fade if I get caught adrift somewhere. There's a backup battery supply, my laptop, and a mobile wireless hotspot to link it to when Wi-Fi is out of reach and I need to get a script approved. I have two IFBs (earpieces back to the control room)—one molded to my ear, another, belonging to someone else's ear, that wound up with my stuff and I hold on to in case mine breaks (as they so often do). And then there's my leather journal, passport, and a couple hundred dollars in small bills that I keep right next to a hidden stash of guilty cigarettes. Longer reporting assignments—like this one—also call for a go-bag stuffed with a few days' worth of clothing, and in this case, a sturdy pair of rain boots.

Now, it needs to be said, at this particular moment in my career I was far from being launched around the world. It was October 2012 and I was a local news reporter for WNBC in New York City. Jamaica, Queens, was as exotic as life got. I've always been a big believer in planning for the future, and on this morning, it arrived. Through darkness, I reached into the main compartment of my backpack down to the dusty bottom where, mixed in with loose change and a stray Tic Tac or two, I found a headlamp that, until now, had only been used for camping. I switched it on, plugged my phone into my backup battery supply, and navigated my way to the kitchen, where my photographer Jeff Richardson was grabbing a bottle of water. "Hope you got rest. Long day ahead," he said in his less-is-more style of communicating. We had just rode out the worst of Hurricane Sandy in his home on Long Island, and it was time to load our stuff into the company car and survey the damage. What we saw in the hours and days to come changed me in ways that, even years later and after dozens of storms by different names, still surprise me. Sandy's wake-up call would ring for days, months, and years to come.

Sandy's Sister Is Born

The oil pumpjacks that seesaw along Texas's highways and interstates go by many names: horsehead pump, nodding donkey pump, donkey pumper, rocking horse pump, grasshopper pump, dinosaur pump, and thirsty bird pump . . . to name a few. But as the sun set along Highway 103, coloring hundreds of miles of vast flatness in shades of orange and pink, I saw towering steel giraffes grazing in a Lone Star Serengeti. "Do you have a light?" asked my producer Bill Applegate. I reached for my backpack, the same one from my Sandy days, and pulled out a lighter along with my own reserve of nicotine. We lit our cigarettes, raised the volume on Springsteen, and lowered the windows. The air, filtered through tall grass, was sweet. It was the summer of 2020, and we were racing for Louisiana ahead of Hurricane Laura's arrival.

Bill Applegate's appetite for nicotine was surpassed only by his craving for gummy bears, indie music, and chasing down story leads. If you were to visit CBS's West Coast bureau while Bill was there, you'd wonder about the then-forty-seven-year-old wilted-flower-of-a-person staring blankly at a computer screen in a corner. He was not a bird for cages. But when breaking news happened, he was the first to raise his hand to go, and once in the field, his humanity was unrivaled. "We're witnessing history," he'd often say. His drive to not only witness history, but also get its first draft right, often meant going a full day without having a proper meal—his shaky hands instead satisfying his appetite with a Camel or Marlboro or whatever brand was available. His father was a titan of local news stations in television's glory days. There was no doubt Bill loved journalism—we spent many nights drinking martinis and talking about what was right, wrong, and plain-old fucked-up with the industry—but I also knew his journey, like so many, was fueled by a desire to make his father proud.

Bill and I covered dozens of extreme weather events together for CBS News. In fact, we were reporting on a wildfire in Central California when I got a call from our national editor: "How quickly can you get to Louisiana?" I'd later come to learn that Terri Stewart's smooth, calming tone was a technique intended to lessen the blow of an assignment that was sure to uproot my life. Terri was CBS's commanding officer of news. To hear her on the other end of the line, and not a regional bureau chief, meant she was about to pitch a priority assignment for me to cover, and that it was capital "B" Big. And also that the answer in response, no matter what was about to come next, was yes. "Looks like Laura could make landfall as a Cat 5," she said. "Probably best you fly into Texas and drive."

We must have passed thousands of those pumpjacks as we made our way from Austin. It took five hours to reach the Louisiana border and another thirty minutes to arrive at Cypress Bend Resort in the town of Many, our base camp for storm coverage prearranged for us by the Southern Bureau. Cypress Bend got its name from the park, which in turn got its name from the bald cypress, a tree that seemed to thrive everywhere in the state—even at the bottom of lakes, where its amphibious roots support thick, craggy trunks that extend more than a hundred feet into the swampy sky. Hundreds of them lined the one-way-in-one-way-out, miles-long road that led to our accommodations. Their long and knotted branches intertwined, creating an archway over our path, and long dangling tendrils of leaves and Spanish moss swayed in the wind. It was beautiful, I thought, before questioning the logic of staying down a path that could easily be cut off.

The power was still on when we pulled up to the resort's main entrance around midnight. There were state utility crews all around, and a few other guys standing outside smoking to calm their nerves. As I grabbed my bags out of the trunk, I overheard one worker say the town didn't have enough lodging for the hundreds of crews on call for post-hurricane cleanup. I got my keys from the front desk and made my way to my room. I placed my backpack on the side of the bed where my husband normally sleeps, set my wallet and watch on the nightstand, and tried to prepare myself for the late night ahead. Rain boots, waterproof pants, jacket, and phone set to wake me at two a.m.

The wind began to whistle through the AC wall unit, and as I looked at the ceiling, preparing for a quick nap in the same smoky clothes I had just worn on a wildfire assignment in California, a big patch of mold caught my eye—then began to move, slowly. It wasn't mold after all—a relief, I guess. Lovebugs. Hundreds of them had broken through the air conditioner to ride out the storm, huddled together. Their wings, overlapping each other, quivered. Leaves in a gentle breeze. I turned out the lights, comforted to have company, any company, in the loneliness of this wheezing room.

Life Before

Hurricane Laura had already knocked out power in the city of Leesville to our south, but the Miller home was electrified with the familiar nervous energy sparked by a looming storm. Fourteen-year-old Cindy grabbed her father's old plastic Duracell flashlight and used her thumb to find the familiar serrated texture of the power switch. She didn't know how old the batteries were or how much time she had left before they ran out, but she cherished the light in this moment and the words it would soon illuminate.

Earlier that day she had been grateful for a different kind of light. The crickety, frog-throated corner of the Louisiana bayou where she lived with her parents and two sisters was home to a sparkle of fireflies. Their ancestors used to light up the night like fireworks, but pesticides and a warmer, more humid climate thinned them out over the generations, and it was especially rare to see any so late in the summer. On this evening, though, a dozen or so twinkled, unaware, or in defiance, of the other act of nature to soon follow. Cindy paused to admire this midsummer mating call. Not many teenagers knew oxygen combined with calcium, adenosine triphosphate, and a bioluminescent enzyme called luciferase produced light, but she did. The pages on bioluminescence in her science textbook had weathered to a linen-like texture after being thumbed through so many times. She was surprised to learn that bioluminescent animals weren't so rare after all, including 76 percent of all marine life. But she especially liked fireflies. While they were usually swatted at like roaches during the day, dusk triggered a chemical reaction that magically turned a bug into lightning and human disdain into awe.

Perhaps Cindy saw a little bit of herself in the fireflies. At just over five feet tall, she almost disappeared into a room if you weren't looking closely. But to find her and to lock your eyes with hers, framed by delicate glasses, was not just to feel seen, but perhaps even understood. She had a weightless and comforting presence about her. She'd been called a "perceptive soul." It was a superpower of hers, honed by years of being bullied by a few classmates who mistook her placid nature for weakness. But this pixie was a *Midsummer Night*'s Puck . . . clever, sometimes mischievous, but always good-hearted. Once in elementary school, after she and a few friends were picked on by another girl, she created a school initiative against bullying. She named it, established criteria for it, got approval and support from teachers, then formed a student committee and organized weekly meetings with the principal. She did it all on her own without help from her parents, at an age where such acts are genuine and pure, not the kind of self-serving performances that develop with adulthood. It was around this time people started to attach "gifted" to her name, which naturally made her the target of even more bullying in her middle school years. But she also earned another adjective, of which she was proudest: fearless. Maybe this is why she had recently cropped her long flowing brown hair. She wasn't one to hide behind drapes. And now, with the hair out of her eyes, she made it a point to see the magic—and the pain—in others. These days, there was a lot of pain to go around.

Living in this part of the bayou wasn't easy. Money was tight for many families. According to the U.S. Census, the poverty rate was 33 percent, nearly three times the national average, which made Leesville one of the most impoverished places in a state that ranked poorest in the country behind only New Mexico. Bravado became its own currency in the bayou. But hurricane season, growing more volatile and erratic in Cindy's fourteen years of life, exposed the real haves from the have-nots. When the power cut out that night, Cindy could hear the distant humming. The vibration of a generator. The sound of money. She had recently read *The Great Gatsby* and had been given a new filter through which to see the world. That diesel-fed engine was her green light at the end of the dock—where the Daisys and Toms of her life lived.

While most of Leesville was stocking up on supplies ahead of the storm, Cindy was worried about missing classes if roads were cut off by fallen trees and power lines. And what if her school got damaged? She may have been just a freshman and only a few weeks into the semester, but she had already made an impression. She lived for getting new assignments and connecting with classmates and teachers. Her perfect grades from middle school landed her in Hicks High's advanced classes, and her teachers would often keep her after lessons to get her take on the curriculum and how their work was connecting with her classmates. In her school's halls, she understood and was understood. She dreamed of becoming a microbiologist and studying bacterial pathogenesis and immunology. When Covid-19 shut the world down, her brain went into overdrive, consuming every article and study she could find. Ultimately, fundamentally, she was driven by a love for helping people, and when she closed her eyes and thought about love, what love really looked like, she always saw her parents first.

Cindy loved her family, even if at times they found her intellectual appetite exhausting or mistook her desire to one day leave Leesville behind as ungrateful. Her parents—James, a pipe fitter out of local 286, and Mary, a nurse at the nearby hospital—made the most of what little they had, and Cindy and her sisters, Nellie and Maddie, always felt protected and provided for. Together they lived in a small, two-bedroom home shielded from the sun by an old oak on a piece of property shared with her aunt and uncle, who lived in a home of their own.

As the storm approached, the family angst that came from being forced to decide when, where, and how to endure the torrent sunk in. Cindy knew the signs. The lines on her mother's face and the ripples in her cup of tea grew deeper. The chatter softened. On this night, Cindy and her sisters piled into her parents' bedroom like quivering lovebugs. It was tight and humid, and now the voices, just like the power, shut off as Hurricane Laura gained ground. The family may not have had a lot of money, but they had each other. Cindy aimed the dusty beam of her father's flashlight on the pages of the latest book from the Percy Jackson series. Soon the threat of the hurricane and the worry of her batteries running out faded.

A Nursery for Monsters

Ten days earlier and half a world away, the sky fogged, then sweat, before pouring out all its contents. The white sand beach, freshly groomed by the receding tide, was now riddled by the rain. The handful of (mainly European) tourists that had traveled here, wondering why airfare and hotels were so cheap, were scurrying off the sand baptized by what locals called *estação das monções*. Monsoon season.

Cape Verde is known as the African Caribbean because of its florescent blue water, tropical climate, and location in the Atlantic just off the coast of Mauritania and Senegal. The island nation gained its independence from Portugal in 1975, but the influence of its former colonizer is everywhere. Cobblestoned streets are frosted with late Gothic buildings in shades of yellow and red and purple and blue that resemble delicate pastries in a glass display at the kind of shop that charges ten bucks a pop. Then there are the actual Portuguese bakeries famed in these streets for their pastéis de nata, bite-sized crispy pastries filled with creamy custard.

Some say it's time to leave Cape Verde when you've gotten sick of eating a nata in the morning . . . or of afternoon strolls listening to street performers sing morna . . . or of ordering cachupa, a warming slow-cooked stew, after a dip in the ocean . . . or of drinking grogue, a style of rum, for a nightcap. Some never do leave. Visitors today marvel at how this absolute gem of a destination remains one of the world's best-kept secrets. And I don't use "best kept" lightly. In 2015, only around 7,000 of the 190 million–plus Americans touring the world touched down here, according to the Instituto Nacional de Estatistica, or the National Institute of Statistics.

The visitors that do come to swim here and snorkel here and sunbathe here and numb the pain of a sunburn with a caipirinha at a sandy bar here often are unaware of the island nation's role as a central hub for the transatlantic slave trade in the sixteenth century. Hundreds of thousands of slaves, the majority from West Africa, passed through Cape Verde. It's not the sort of thing a budding young holiday destination advertises. Better known is Cape Verde's penchant for shipping hurricanes to the Americas. Every year during a three-month stretch between July and October, the spicy, hot, dry air of the Sahara Desert in the north clashes with the earthy, damp air from the south, ushering in estação das monções and triggering, much higher up in the atmosphere, powerful winds known as the African easterly jet. It's a high-altitude alchemy as old as time, but our warming atmosphere is warming our oceans, which in return are releasing more water vapor into the air, providing the kind of fuel that can quickly turn a tropical wave into a tropical depression, then a tropical storm, before becoming monsters with familiar names like Andrew, Isaac, Frances, Bertha, Florence, and Irma. According to NOAA, 80 to 85 percent of all major Category 3, 4, and 5 hurricanes are birthed around Cape Verde and then make the 2,000-plus-mile journey west, gaining more and more devastating power from the warmth of the ocean with each nautical mile.

On August 16, 2020, as those sunbathers scurried over the blemished white sand, a tropical wave entered the world. It then became a tropical depression on August 20, and a day later earned its name: Tropical Storm Laura. She first hit the Lesser Antilles and brushed Puerto Rico, then moved across the island of Hispaniola and wreaked havoc in Haiti, where she killed thirty-one people, as well as four in neighboring Dominican Republic. Laura then moved across Cuba, leading to the evacuation of 260,000 people there. By then the forecast models were clear as to the relative direction she would take, but even they couldn't predict the rapid spike in intensity that would come.

On August 26, Laura crossed the steaming Gulf of Mexico, a whole ten degrees warmer than usual according to weather reports at the time, and in just twenty-four hours graduated from tropical storm to Category 4 hurricane with sustained winds of 150 miles per hour. Weathercasters on television were stunned. Just hours earlier there was hope the storm wouldn't top a Category 2. Panicked people in the Gulf Coast states of Texas, Louisiana, Mississippi, and Florida flooded grocery stores, waited in hours-long lines for gas, and drove around in search of the rare ATM that still had cash. The rapid intensity of the storm unnerved even the most weathered preppers. Beachside businesses boarded up their windows with sheets of old plywood spray painted with messages like "Leave Us Alone, Laura"—and the crossed-out names of the previous storms the same wood had been used to defend against. One went all the way back to Katrina. As Laura got closer and stronger, hurricane warnings and watches turned into mandatory evacuations in parts of Louisiana. The town of Leesville was not one of them.

Homecoming

It was four in the morning. The gusts of wind that had first chimed through the neighborhood were now squealing. It was the kind of high-pitched sound that produced sparks in the brain like a saw cutting through steel. The metallic chords would later be described as a freight train. A very long freight train. It was ultimately the noise of resistance. Fences, a basketball hoop, an aluminum-roofed carport, and trees were standing their ground, but each gust loosened their bolts and nails and roots, and it was anyone's guess how much longer they could hold on to the earth.

Cindy was wide awake, still reading. The batteries of her flashlight ran out sometime after midnight, but her plan B, a small LED mirror her mother used for applying makeup, was working just fine. Her parents and her two sisters were all awake as well, though Nellie, the nineteen-year-old, began to whisper as her parents' responses grew more and more delayed. The weight of sleep was fast approaching, even as the storm barreled down.

They had lost track of Hurricane Laura's path hours earlier after it had made landfall 120 miles south. Last they heard, wind and flooding were the biggest threats. But without power, they couldn't watch TV or listen to the radio for updates, and cell service was spotty at best. They were comforted by the fact that Leesville wasn't on any evacuation list. Others would wonder if they were forgotten.

One meteorologist later described the modern phenomena that unfolded as Laura intensified over the gulf, and the destruction to soon follow, as Mother Nature's "nuclear meltdown." What glaciologist John Mercer first warned about decades earlier, and what researchers who followed in his footsteps were witnessing in another remote corner of the world, had arrived on American shores.

Answers on Ice

Icebergs the size of Miami Beach condos sailed by as sunrise turned this polar paradise pink. There are a few places on Earth that, no matter where I am or what I'm doing, I can close my eyes, think about them, and instantly be transported back to that moment of discovery: my father's woodshop tucked away in the basement of my childhood home; the tidal pools of Eastham on Cape Cod; the argan forest outside Marrakesh; Colorado's Maroon Bells during autumn; that bench near that bridge in Central Park; and now the Sermilik Fjord in Eastern Greenland.

Those icebergs showered me and the fire-engine red vessel I was on in refracted disco sunlight. The icy mist glittered. It was hard to believe that just two months before, this reality wasn't even a thought when I picked up the phone and cold-called a physical oceanographer with the Woods Hole Oceanographic Institution to ask about her research on the rate of Arctic ice melt. "I'm afraid a spreadsheet of water temperatures isn't the most visually appealing for your television cameras," Fiamma Straneo warned with a soft Italian accent. "But my team is headed to Greenland to collect new readings near a glacier, and that's exciting. What are you doing in August?" It sounded like a dare.

It took two days and four flights to meet up with Fiamma's team in Tasiilaq, a small east coast fishing village sprinkled with vibrantly colored huts that looked like the homes on a Monopoly board. A small bar advertised imported beer and salted cod. It was late summer and most of the snow and ice had melted on land, revealing the Danish territory of Greenland, the world's largest island, is in fact green under all that winter white. But winter, I soon learned, was easily found on the frigid water.

The fjord, an arctic river linking Greenland's primarily land-based ice sheet to the ocean, was so calm it looked frozen, and I could read the name of our boat in its reflection. The *Viking Madalex*. It was 2013, and Fiamma and her team were researching scientific threads dangling from glaciologist John Mercer's incomplete work. Funding for this modern expedition was tight, which meant a small fishing boat converted into a research vessel would have to do. The fish guts and sailor talk had been wiped clean the previous night and you could have almost been fooled into thinking you were on a world-class floating laboratory had it not been for the captain, a gray-bearded, salty man who always had a cigarette dangling from his chapped lips. He was carefully guiding *Viking Madalex* through the Sermilik Fjord to the eastern wall of Greenland's ice sheet.

"Ice sheet" is a deceiving term because it is not one solid mass of ice. It may be easier to picture it as a massive tray of tightly packed ice cubes that shifts and even flows under the pressure of its own weight. Greenland's ice sheet is two miles thick and covers an area the size of Mexico. It's the second largest sheet behind Antarctica. While most of it is on land, its outer frozen margins terminate in fjords where they appear like ice walls holding the rest of the cubes back, slowly dispensing them into the ecosystem over time.

"It is natural for ice sheets to grow and shrink with the seasons, gaining ice from snowfall in the winter, and shedding, or calving, ice into fjords in the summer," Fiamma explained. Ice sheets remain stable as long as they accumulate the same mass of winter snow as they lose to the sea in the summer, but Greenland had been steadily losing ice mass since the '70s and was now melting six times faster than it did back then. Greenland's ice sheet was shedding an average of 279 billion metric tons of ice per year, the most out of any ice sheet in the world, contributing to one-quarter of global sea level rise. The eastern ice wall the captain was taking us to, known as the Helheim Glacier, was calving ice at historic rates. Fiamma and her team came all this way to better understand what role warmer water was playing in this rapid acceleration.

The bergs got denser the closer we got to their source, and I sensed the captain preferred the freedom of fishing on the open water over this more constricted science expedition. What he may not have known then was the research he was piloting would later help explain why his daily haul of fish was changing in species and dwindling in size. For now, though, his eyes were on the tips of the icebergs and what lay below. Icebergs go by different names based on their shape, size, and color. There's glacier bergs, named for their irregular shape. Tabulars, identified by their flat top. Slopers, because of their slide-like appearance. Dry-docks, a kind of berg eroded in the middle, giving the appearance of twin bergs attached at the hip. Blockys because they look like blocks. Weathered because they're . . . weathered. And my favorite, pinnacled bergs, whose sun-catching spires resembled frozen fire in this arctic snow globe.

The research team of about ten had joined me on the bow to take photos of these floating cathedrals. Fiamma was there, too, equally awestruck. "Scientists helped bring awareness to Antarctica but were not able to predict the kind of rapid loss of Greenland's ice we're seeing today," Fiamma said. "It's exposed gaps in our understanding of how oceans and glaciers interact and what that interaction means for our future climate." She wrapped her small frame in a black down jacket and wore a red knit hat over her short black hair. Reading glasses dangled around her neck. She looked like a grad student, until the ship slowed down, and it was time for her work to start.

"I think we're over it now," she said as her team snapped into their roles, preparing to haul in the science experiment that rested on the fjord's floor. Everyone was quiet as they powered on electronics and laid out ropes. No movement was wasted. The laughter and excitement from earlier had been sharpened into a razor-like precision. To make a mistake now, to return home empty-handed, would compromise critical work and possibly future missions. I watched the captain as he quietly observed. I thought for sure he'd break the tension with laughter or a glance at his crew. But his face looked just as tense as those of his adopted team, as they, like all fishermen, paid respect to the search for life in a deep abyss.

That life Fiamma was in search of was one of several sensors the size of a scuba diving oxygen tank that her team had moored to the bottom of the fjord a year earlier. It measured water temperature and salinity every hour of every day. When plugged into a computer on board *Viking Madalex*, Fiamma could see how much the ocean temperature had risen. Years of collecting and comparing this data would reveal a pattern. Engineer Will Ostrom was in charge of extracting the device. "It's tricky. There are more icebergs in the fjord this year," he explained as he scanned the glass surface of the fjord. "The instruments run the risk of getting stuck beneath them." A yellow box by his feet was emitting a tone that sounded like dial-up internet from the '90s. This Morse code, or something like it, was telling a latch connecting the sensor to the mooring below to let go. If all went to plan, a buoy attached to the censor would rocket the time capsule to the surface for collection. Within minutes, we heard the splash.

Will, Fiamma, and the team now chirped with excitement. Their prize catch had just breached, and they were about to bring it on board in a process that looked to my novice eye like a life-sized arcade game. The captain, fresh cigarette dangling from still-chapped lips, now pulled on levers that spun a crane equipped with a steel cable and hook into action over the ship's stern. Using a pole, Will latched the hook onto the sensor. "We got it!" "Okay." "Bring it in." "Slowly." The captain pulled another series of levers and hauled the sensor into captivity as the team looked on. The crisp water from the fjord now showered down on *Viking Madalex*'s deck. "That's awesome." "And on the first go."

There was little time for celebration. It was the first of many devices that would need to be retrieved and replaced over the course of the week. The team would also be deploying new measurement tools, including an autonomous, remote-controlled kayak rigged with water sensors. This allowed them to safely get the most accurate readings as close as possible to the crumbling wall of ice under threat.

The expedition was part of a larger collaboration with another team collecting data from a different perspective. That's how, days later, I found myself leaning outside the open door of a twin-engine helicopter, taking photos of Helheim Glacier's fractured surface below. The inky lines in the ice created a pattern that resembled a smudged fingerprint. "What you're seeing there are crevasses essentially carved into the ice. Some are hundreds of feet deep," glaciologist Dr. Gordon Hamilton told me through the headsets covering our ears. "They're created when the ice moves and likely amplified by climate change." I could see my reflection in his sunglasses, which briefly distracted me from thinking he resembled a pilot from *Top Gun* more than he did a scientist at the University of Maine. He was taking me to the base camp his team had set up overlooking Helheim's mouth. The data they were seeing at this field lab revealed rapid change. "My colleague almost fell out of his chair. It's that alarming," he said as we approached. He spoke in clear, troublesome terms like I imagined John Mercer did before him.

The helicopter dropped us on a rocky perch a short walk from where the team's tents were pitched. As you now know, I have a knee-shaking, palm-sweating fear of heights but was immediately drawn to the edge. The ice, its texture, looked like a stormy ocean had suddenly frozen over. "Do you see the line on the rock wall? It kind of looks like a bathtub ring," Gordon described as he waited for my eyes to lock on to what he was pointing at. "That's where the surface of the ice sheet used to reach."

After recently analyzing a combination of satellite imagery and data from his team's ice penetrating radar, Gordon was able to determine Helheim had thinned by more than 300 feet in just over a decade, like a massive soufflé that had been deflated. In the same period, the glacier had retreated by more than 4 miles. His research also included dropping GPS devices from the open door of the chopper onto the glacier to track how fast the tray of ice cubes was dispensed into the fjord, causing that retreat. Those trackers revealed one part of Greenland's ice sheet was moving 8.7 miles a year, meaning icebergs were being dumped into fjords at more than twice the rate they were in the '90s and the ice sheet wasn't gaining enough winter snowfall to rebuild what was lost. "Humans have never witnessed this scale of loss before, and not enough people are as worried as they should be," he said, sounding now like NASA's James Hansen. His sudden vulnerability and frustration was unexpected for a man burnished by adventure. While he lived a life worthy of a *National Geographic* cover, the risks he and his colleagues took were ultimately in pursuit of the truth, which only really mattered if people listened.

Together, Gordon and Fiamma's findings helped expose the root cause for Greenland's rapid thawing, and also proved to be missing pieces to another puzzle that would connect what was happening in Greenland to extreme weather in other parts of the world. The early data collected by Fiamma's sensors revealed average water temperatures of 40 degrees—warm enough to rapidly melt Helheim. The ice sheet's eastern wall was developing major cracks. "We've recorded 39 degrees on previous trips, but we've never seen water this warm in any of the fjords in Greenland," Fiamma told me. "Average air temperatures are rising here faster than anywhere else on the planet, and the impact of that warm air has long been connected to accelerated melting, but what we're seeing here is another threat that's never been factored in because the methods of measuring it didn't exist before," she said. Warm air and warm water were burning the ice on both ends. And with the base of Helheim lubed with warm water, ice easily slid into the fjord. "We're witnessing the possible collapse of a critical regulator for our planet," Gordon said. That collapse wasn't just impacting sea level rise.

The Greenland ice sheet plays several critical roles in stabilizing our planet's climate and nourishing our habitat. Like a mirror, it reflects the sun's energy back to outer space, a reflective shield that decreases the amount of heat absorbed by our planet. Less ice cover means more heat, means more melting, means less ice cover. It's a dastardly loop. Then there's the impact from greenhouse gasses, which prevent heat from escaping. Our oceans have absorbed 90 percent of excess heat trapped in the atmosphere because of greenhouse gasses since the 1970s, as recorded by NOAA. The ocean's surface layer, the top 700 meters where most marine life live, takes most of that heat and has warmed an average of 1.5 degrees Fahrenheit since 1901. More heat means less oxygen, means less aquatic life that can survive. In the arctic, research like Fiamma's has helped reveal that ocean temperatures in that region have risen on average by more than 3 degrees. It's a fever that won't break in an ecosystem that can't afford to call in sick.

The ice sheet also functions as a frozen freshwater reservoir, collecting and dispersing water as part of a larger cycle that helps regulate ocean temperature and salinity, like a heart pumping blood through veins, as Professor Vaughan Lewis once described. Today, thanks in part to research led by Gordon, we know this Arctic organ is bleeding out and the unprecedented flood of freshwater that's entering the ocean is diluting the Atlantic's salinity, reducing its density, and in turn slowing down a conveyor belt of underwater currents that circulate colder arctic water and vital nutrients to the warm south. All that warm stagnant southern water evaporates quicker, feeding moisture and energy into storms. In other words, left unchilled, our warming Atlantic is the perfect fuel for hurricanes. According to a 2023 study published in the journal *Nature*, the rapid melting of our freshwater ice sheets could shut down this critical conveyor belt by 2050.

Another study by NOAA used satellite images to link warming in our oceans to intensifying storms. Reviewing satellite images dating back to 1979, researchers found warming had increased the likelihood of a storm developing into a major hurricane by about 8 percent a decade, or 40 percent between 1980 and 2020. All of this coincided with Greenland and Antarctica's rapid melting. Hurricanes were intensifying quickly, much like a nuclear reactor melting down when its coolant is removed.

I left Fiamma's and Gordon's expeditions early so I could continue my reporting in another part of the country. A fishing village up north was under attack by starving polar bears who were running out of ice to hunt seals. I spent my last day in Tasiilaq on land as Fiamma and Gordon continued their work at the ice sheet's edge. At dusk I walked around the village taking photos of those fishing huts and the countless sled dogs resting outside ahead of a busy winter season. I was seeing the village with fresh eyes. This remote fishing outpost, and the icebergs that passed through its backyard before entering the Atlantic, were really the gateway to the rest of the world. In the silence I heard what sounded like small avalanches breaking the trunks of trees. Helheim was releasing more icebergs into the fjord, where they crackled and fizzed in this glass of warming water.

I wondered if our global response to climate change would have accelerated had John's work been celebrated and expeditions like Fiamma and Gordon's happened decades earlier. I also thought about the risks they took to sound the alarm that many ignored. It must be hard to see something that no one else can or will and to work in a profession that focuses on the future when most can't focus beyond today.

Dr. Gordon Stuart Hamilton died tragically in 2016, while studying ice loss in Antarctica. A thin layer of ice collapsed under the weight of the snowmobile he was on and he plunged into a crevasse. He was fifty years old. In remarks made following his tragic death, colleagues found solace in knowing he lived long enough to witness world leaders finally take research like his seriously. He called the 2015–16 Paris Agreement treaty on climate change a small victory. But he also knew it was only the beginning. Progress would need to be swift to protect communities around the world.

"No Man's Land"

Follow First Street past the office of Wayne Bush Attorney at Law and Marshall's Spin Cycle Laundry until you hit Lula Street. Here, just a few doors down from the local Probation and Parole Division, you'll find Leesville's oldest building—a single-story plantation-style house erected in 1855 by slaveowner Dr. Edmund Smart. In 1871, after the South lost the civil war to the North, Dr. Smart and his father, John, established Leesville in honor of Confederate general Robert Edward Lee. Modern Leesville has evolved since then, but it remains a divisive, yet fitting name for a community on the front line of several turbulent periods in American history. Leesville, population 6,000, has witnessed more change in the past two centuries than many major American cities ever will.

After the 1803 Louisiana Purchase made the U.S. and Spain neighbors on the continent, a land dispute erupted over a sliver of land on the southwestern border between the newly minted state of Louisiana and then-Spanish Texas. The two countries agreed to turn the region into a neutral zone appropriately named at the time "No Man's Land." Any signs of previous ownership were quickly reclaimed and covered by the dense marshy forest, turning it into the perfect place for outlaws—cattle rustlers, bandits, and the like—to come and hide.

This wild South also attracted escaped slaves who had mainly been brought to the region during French rule to work the area's plantations. Their journey to freedom along the Mexico-bound branch of the Underground Railroad was guided mostly by word of mouth, and they escaped slave hunters and their bloodhounds by hiding in a network of safe houses. Their target was the nearby Sabine River, a 360-mile-long waterway that snaked downstream toward the gulf. Over just a few decades, Spanish-owned Texas became Mexican-owned Texas, which then became the independent Republic of Texas before Texas became the twenty-eighth American state, and one of thirteen that demanded legalized slavery. An untold number of slaves crossed the Sabine in No Man's Land to eventually find freedom in Mexico before slavery was de jure abolished in 1860.

In the years that followed, pioneers flooded the swampy outpost, then officially recognized as Leesville, to mine the same haunted forest for its lumber until most of the oaks and cypress, along with much of their history, were clear-cut in the 1930s. In 1941, the United States

Army built Fort Polk just south of town and opened a prisoner of war camp for Germans captured in World War II, who were then sent to work in the muggy fields picking everything from cotton to rice. The Army gave former No Man's Land a new nickname in the 1960s: Tiger Land, because the jungle-like environment was considered ideal for preparing troops to fight in Vietnam. More than one million soldiers trained in two simulated Vietnamese villages that featured earthen berms, sharpened bamboo stake defenses, and booby traps. A sign at the entrance to one of the thatched villages welcomed trainees to "Tiger Land—Home of the Vietnam Combat Soldiers." It was like Epcot Center for warfare. By 1969, Fort Polk had dispatched more soldiers to Vietnam than any other military post in the country.

Leesville's economy has ebbed and flowed with the changing tides of war, with many of those in town servicing and supporting operations on base. Today the Sabine River is enjoyed in the spring for bass fishing, and feared in the summer when hurricanes have been known to push a wall of warm salty water from the gulf upstream and into communities like a backed-up sewer. Flooding has increased fivefold since 2000, according to NOAA, and rising temperatures have been linked to hundreds of heat-related injuries and even deaths at Fort Polk. In an eight-month period in 2020, two hurricanes, a tropical storm, and an ice storm struck the region, collectively damaging tens of thousands of homes and causing extensive damage to more than 70 percent of military housing on Fort Polk, now known as Fort Johnson, reported Corvias, the company contracted by the U.S. military to manage the properties. Climate change is now the latest battle being fought and fled from in this corner of America's Deep South.

Follow Holly Grove Road outside Leesville's borders, through thick new-growth forest, until you reach Clemens Road. Here, a few yards away from the cemetery's porous headstones and a two-story white clapboard church, you'll find a metal plaque staked in grass that reads: "First Church services 1826 . . . Never without a pastor. Still on original location." The Holly Grove Methodist Church played a critical role in housing escaped slaves before they used a popular trading route to reach freedom. Today, El Camino Real de los Tejas, or King's Highway, is recognized as a national historic trail, and every hurricane season, people follow a similar route as they flee to higher ground.

Sitting Ducks

The heat woke me up before my alarm. Laura's outer bands were approaching this stretch of King's Highway and had knocked out the power. The air in the room was moist and thick, and the sound of wind and rain filtered through the now-useless wall-mounted AC like the static of FM radio while driving through a tunnel. I put my watch on, stuffed my wallet into the pocket of the pants I napped in, put my waterproof pants on over them, pulled on my boots, grabbed my backpack, and reached into the front pocket for my IFB. I placed it in my ear, threw on my raincoat, grabbed a Red Bull out of the side pocket of my backpack, tossed the bag over my shoulder, and headed out the door. All in under five minutes. Not bad. I met Bill and the rest of my team at our live shot location just outside the front entrance I had arrived at hours earlier.

"We're the top of the show," Bill said while lighting a cigarette. "Laura will be passing over us during air, so it's going to be a long morning of updates." It's at this point of every hurricane, the ten minutes before that first live shot, where I go through what I'm about to say to make sure whatever I describe is accurate and whatever I show viewers doesn't make me look foolish. Every so often, especially while covering a hurricane, I'll read a Tweet from a viewer questioning the logic of having a reporter go live in the middle of a storm. It's fair criticism, because unlike wildfires, where our team watches Earth scorch from the sidelines, in hurricanes we go right into the eye. But what many people don't realize is that networks like CBS invest a lot of time and money into training their field teams to know how to cover everything from wars to riots to, yes, extreme weather safely.

I don't think viewers are actually turned off so much by the risks that come with storm coverage as they are the theatrical razzle-dazzle some journalists add to their reporting. Using rope to tie yourself to a sturdy object so you don't blow away in the wind, for example, is probably not a good way to earn trust. Personally, I'd rather blow away in the wind than go live to America while rocking back and forth talking about how the wind is throwing me off-balance as folks casually stroll in the background. And if the reporter on air is wearing a helmet and safety goggles, as if those things could really save them from an object tossed in 100-mile-an-hour wind, I've got advice for you: Switch the channel immediately.

There's been several times when my team has decided it wasn't safe to report. Today wasn't one of those days. Our live shot was set under the sturdy overhang of the hotel and the worst of the wind set to hit Leesville was still an hour away. And so I stood my ground firmly, no tethering rope in sight, and composed myself as I waited for my cue.

"Good morning, Gayle." I slipped into reporter mode, speaking through the worsening storm around us. Going live in the path of an approaching hurricane can be an unnerving experience. The power of Mother Nature is one thing, but then there's the toll she collects before letting the sun come out. It's hard not to think of the real lives threatened as I report on updated warnings and expanding evacuations, all the while shielded by the kind of security and privilege only a network news organization can buy. As I report with little fear, I'm always aware of the countless people who are not only terrified, but in many cases unable to find or afford safe shelter. In these adrenaline-fueled moments, every word feels like a self-fulfilling prophesy—a curse—waiting to find a sitting duck.

I did ten reports that morning. Each "hit" featured new information being passed along to me by Bill and my producers in New York, who were receiving updates and flash flood alerts from emergency response agencies. By our last round of updates, around ten a.m. Eastern, we were hearing of water rescues and the growing threat of tornados. One person was trapped by a fallen tree and emergency crews were struggling to get through blocked roads.

"What's Your Emergency?"

"Cindy? Cindy, are you okay?!"

There was no response. A mature oak tree crashed through the roof and swallowed the corner where Cindy had been reading moments before. The leaves streamed in the wind like thousands of tattered flags, half-raised. As her father, James, searched for his daughter, her mother, Mary, called for help.

"Nine-one-one. What's your emergency?" The operator's calm and measured voice, a tone softened by years of riding the blinking lines, sparked a moment of hope as Mary struggled to find words to describe the scene in front of her. To say out loud what she was witnessing was to actively test her faith. Cindy was trapped under a fallen tree and unresponsive to their calls, she finally explained. James struggled to lift the oak, but it was too heavy. "Is she okay, James? Is she okay?" asked Mary. Her words came out in a pitched singsong that rose and fell between hope and despair. It was a sound the operator had heard before. The landsliding lullaby of a mother soothing her broken child.

Help, the operator told Mary, was on the way, but trees were down across the region and it could take a while. James had taught his daughters to never give up. There was always a solution. You just need to take a moment and think, he would tell them. He was now taking his own advice. He asked Nellie and Maddie to help their mother comfort Cindy as he climbed through the shredded wall of their home and into the shredded backyard on the other side. Hundreds of splintered trees blanketed the bayou like barbed wire and the road was blocked, preventing him from getting out and help from getting in.

Residents in Leesville didn't hear many sirens that morning after Laura passed. Instead, they recalled that metal melody of chainsaws . . . a twisted tune distorted by miles of reverberation through a now-unrecognizable landscape. Two bands of men, cut off by more than a dozen massive trees, were chiseling away at their ends. First responders including sheriff's deputies and volunteer firefighters were on one side of the road. One mile away on the other side, James and his neighbors were at work. And James knew they had their work cut out for them. He did the rough math. It was taking about thirty minutes to clear each tree. He stayed focused and thought about the summer afternoons he and Cindy spent listening to old-school rock and roll while restoring his vintage Camaro. He had promised her she could drive it to school once they got it up and running again. As sawdust collected on his beard, James thought about his wife and children and prayed to God for a miracle.

Notes from a Post-storm Philharmonic

How many sounds does New York City make? Like, if you were to stand on the corner of Clinton and Houston in Manhattan, or Webster Avenue and Fordham Road in the Bronx, and record for an hour, how many different sounds would you be able to identify? There's the moaning and screeching of a city bus, the cha-chung-cha-chung-cha-chung of the subway passing under your feet, the beeping of a truck backing up, honking, construction, a skateboard, the jingling of a dog's collar, an ambulance siren, the beating of a basketball. It's an urban philharmonic. A never-ending parade of humanity in a city that never sleeps. But on October 29, 2012, raging water, howling wind, and searing fire reduced parts of this metropolis to barren earth. In the days following Hurricane Sandy, New York City was trapped in a spectral slumber.

People talk about the calm *before* the storm, but rarely mention the purgatory that follows when the turmoil ends and conditions are still too dangerous for even emergency services to have made their way to the damage. In the purgatory after Hurricane Sandy, the absence of sound created its own noise. A ringing in the city's ears, like tinnitus. At least, that's what it felt like as photographer Jeff Richardson and I approached Beach Channel Drive and 35th Street in Far Rockaway, Queens, to survey the hurricane's aftermath. Jeff usually wore a blue-collar mix of tan Carhartt jacket, stonewashed blue jeans, and Timberland boots while out on the streets filming, aspiring to something like a Diane Arbus–eye on street life but adorned with his high-and-tight flattop. In a city oversaturated by Sandy's chaos, Jeff's knack for spotting moments others overlooked would come in handy.

First we saw the massive metal shipping containers tossed like toys, blocking roads and leaning up against buildings and traffic lights, their origins unclear. "Someone must be hurt," I said to Jeff, pointing at an ambulance stopped in the middle of the intersection.

We flipped on our hazards and inspected the scene on foot, but there was no sign of anyone. I approached the ambulance cabin to find it abandoned and full of water. A walkie-talkie was off its receiver and sitting on the driver's seat, next to a clipboard of paperwork bleeding ink. This was a $500,000 vehicle—more expensive than a Lamborghini—abandoned in the middle of the street with its keys still in the ignition.

"That's the water line," Jeff called out, pointing to a brown line that ran the full length of the vehicle, just above our heads. Above it, the ambulance was clean. Below, it was stained and plastered with leaves and other debris. That same horizontal line divided all of Far Rockaway for as far as we could see. Had we been here when Sandy passed over, we would have been quite literally in over our heads.

A storm surge occurs when high winds from weather systems like hurricanes push large bodies of water toward the coast. In Sandy's case, the tsunami-like phenomenon pushed more than 13 feet of water into parts of New York City, turning a concrete jungle into an urban aquarium. Millions of gallons of seawater flooded the Brooklyn-Battery Tunnel from floor to ceiling. Emergency teams, including doctors in their scrubs, evacuated 300 people from NYU Langone hospital after floodwater destroyed the electrical system. The water bubbled through sewer grates and into East Village storefronts. And in Far Rockaway, the Atlantic briefly connected with Jamaica Bay, flooding everything in between.

The abandoned ambulance told a story of first responders forced to flee. I was observing a world frozen in time, like a stopped clock buried in the wreckage after a bomb exploded. At least, that's how I described it as we went live for WNBC's morning show. We reported from countless still-lifes that day, like in Gerritsen Beach, where we discovered a single-story clapboard hut about 20-feet long and 10-feet wide resting in the middle of a street. The white windowsills looked freshly painted. Tables and chairs were still inside along with dozens of bottles of liquor and a full jar of bright red maraschino cherries. The refrigerator was filled with beer, still cold. We were standing in front of what used to be some kind of bar. But where did it come from? There were no skid marks on the ground indicating a direction, and even more surprising, no people milling around. Just empty streets and that ringing sound of silence.

Thirty minutes later a man approached, stumbled really, already talking to me before I was even in earshot ". . . at first there was two inches of water. That's when I put my father in the car. In seconds it was up to my waist. I threw my mother over my shoulders. I was freaking out," Charles Connelly said. He was exhausted, and his jeans were slick from whatever chemicals they had absorbed in the surge.

I was the first person he had seen in hours and he was eager to unload what until now he was left to process alone. "I called nine-one-one but they said there would be no help until after midnight when the water receded, so I just kept making trips. There were so many people trapped in cars. Even after midnight, nine-one-one never showed up. I thought maybe you were them." Another man wandered over, pulled by misery's magnetic force. He had watched from his bedroom window as the bar drifted, in what looked like a river. "It just floated through my backyard at like five miles an hour. It smashed through my gate and over submerged cars."

After I went live, a viewer helped connect the dots. The bar was called The Sugar Bowl, a popular beachside hangout in Breezy Point, a community on the other side of the inlet. Sandy's storm surge lifted the bar off its foundation, and sustained winds of 90 mph pushed it like a sailboat four miles away. But The Sugar Bowl, it turns out, had gotten off easy. The sliver of sand it was plucked from was engulfed in what would become one of the most destructive fires in New York City history.

————

"How quickly can you make it to Breezy Point?" our assignment editor asked as I watched Charles and his neighbor inspect their new neighborhood bar. We drove over the Marine Parkway Bridge, past fishing boats stranded on the road, made a right on Rockaway Point Boulevard, and kept going until we reached the security booth at the main entrance of the gated enclave, which had been taped off by the National Guard. It was the first disaster relief we had seen all morning. There was a woman on the side of the road crying on the phone. "It doesn't look good," I heard her say to the person on the other end. We waited an hour for the remaining floodwaters to recede, and were then escorted in.

Very few New Yorkers have heard of, let alone been to, Breezy Point, but if you've ever strolled along the Coney Island boardwalk, chances are you've seen the rooftops of small beach bungalows nestled on the tip of Rockaway Peninsula. The remote strip of sand in the nation's largest city can only be accessed by a bridge and single road.

Irish firefighters and police officers first settled Breezy in the early 1900s, and the sandy outpost is lovingly called the Irish Riviera by those who live on it. You can still find some of the original beach bungalows, passed down from generation to generation, in the old district, a streetless, car-free triangle of land locals call the Wedge. Homes here are only separated by narrow footpaths. From the window of a plane landing or taking off at nearby JFK, the tightly packed bungalows look like a quilt. A close-knit neighborhood. The population has grown over the decades to about 5,000. Many of the residents here are still firefighters and police officers and remember the day they saw smoke rise from the Twin Towers in the distance and rushed in to help. Breezy lost more residents per capita on 9/11 than any community. Many of them were first responders. There's a memorial for all thirty-two victims on the bayside beach, not far from where The Sugar Bowl served the thirsty. The cross-shaped sculpture, made from steel recovered from the World Trade Center, survived the inferno that rose from the water as Sandy hit.

About a hundred people—those who defied evacuation orders and others who volunteered to safeguard the community—were witnesses that night when Breezy Point became Atlantis. "I was stationed at the security gate and had to abandon it. The water came in like a wall," Tim Beirne told me. He and another volunteer loaded into a pickup truck and set off for higher ground, but it was too late. The streetlights and house lights flickered before going out, like a scene from *Titanic*. "The ocean rose so quickly, water was reaching our windows." The surge lifted an SUV in front of them and it started to float away. Tim's truck remained grounded, weighed down by people who were jumping into the vehicle's flatbed. He radioed fire chief Eddie Valentine for backup, not realizing the chief was dealing with his own crisis. The water had rushed into the volunteer fire department, flooding their two trucks and a pair of ambulances. The town's only means of emergency response was now offline. The force of the storm's surge also dislodged homes from their foundations, shattered porches, and uprooted utility poles. Down the "road," a team searching by borrowed motorboat for trapped people had to turn around because the current was too strong. The jet-black water, jagged with the splintered remains of homes, was sanding down everything in its path. The worst of the torrent lasted for about ten minutes. Then, as quickly as the riptide came, it stopped. The Atlantic had reached the bay. The waterline was now rocking back and forth in an arm wrestle between two typically divided bodies of water. Each drop of rain landed with the weight of a penny, and millions of them pelted the water's surface as Halloween pumpkins, Adirondack chairs, and even a kitchen sink (no joke) bobbed on the surface.

The damp darkness of the Breezy Point community center provided some refuge, but the night was far from over. Looking out a window, a resident noticed bungalows in the Wedge backlit in flickering orange.

The saltwater had come into contact with electrical wiring and sparked a fire that was now spreading rapidly, driven by hurricane-force winds. With Breezy Point's fire trucks water-logged into stasis, bungalow after bungalow caught fire. The inferno raged unopposed for four hours until the wind calmed down.

Only daylight could reveal the true extent of the loss, and Jeff and I both sat silent as we drove through this wrecked ecosystem. The Wedge smoldered for blocks. All that stood were a few chimneys in a salty mist of smoke and sea. The fire took 111 homes down to their cinder-block foundations. Many others were damaged. I found retired firefighter Bob Reilly rustling through the charred remains of his home. "This is worse than I imagined. I mean, everything is completely gone," he said through tears as he pulled out the blackened frame of a bike he used to pedal along the nearby boardwalk. His wife and daughter were hugging in what used to be their living room. "The important thing is we're okay." Down the street from Bob and his family, I met Christina Kirk. She was crying. Her home had sustained water damage but was spared by the fire. "My son put a rosary on the door before evacuating," she told me. "That's the only reason my home is still here today."

Incredibly, there were no deaths or major injuries recorded in town, no doubt thanks to the heroic work of volunteers who brought many of the holdouts to the community center before the fire. But what happened in Breezy Point marked a new understanding and fear of the modern tempests threatening our coasts. Sandy rendered all lifesaving tools known to man useless. Even a town of firefighters in a city like New York, with all the resources in the world, couldn't rebound quickly enough. In the days that followed, Jeff and I witnessed what happens when Mother Nature outpaces human capacity—and willingness—to adapt. We saw hundreds huddled around power strips connected to generators, waiting for hours to charge their phones as the power remained down for days. Neighbors checked on missing neighbors in flood zones. A child handed loaves of bread to an elderly couple who were among dozens of people I watched dig through a dumpster of rotten and expired food tossed out by the local supermarket.

Everyone we met was treading lightly. They were exhausted, for sure. But I also got the sense that, much like me, they were afraid to make noise. Maybe they were worried about being spotted as an outsider on a turf whose lines were no longer clear. Or perhaps they knew all good deeds have an expiration date, and were concerned about being cut off before their turn.

This parade of ghosts could have been prevented had people been evacuated before the storm, but some of New York's hardest-hit neighborhoods weren't even considered at risk of flooding. Hurricane Sandy not only brought the world's greatest city to its knees, it exposed the nation's failure to recognize climate change and habitat change as a serious threat, and their failure to confront this threat head-on and properly protect those in harm's way. History was repeating itself. Had politicians listened to private insurers back in the late 1920s, many of America's most vulnerable areas, including parts of Sandy-ravaged New York City, would have been developed differently, or never at all.

Regulating Sandcastles

Some parts of the Mississippi weren't meant for civilization. It was always a wild, snaking river that would naturally flood the plains around it for miles every spring. But in the sixteenth and seventeenth centuries, French explorers tried to domesticate the mighty Mississippi in a bid to harness its economic potential. Their experiment began with the settlement of New Orleans in the early 1700s when they built the first levee, a system of walls created to hold back the water. This habitat change continued long after France handed over the keys to the United States. With the help of the U.S. Army Corps of Engineers, channels were dredged to make way for ships and more levees were constructed along the river to create space for agriculture and dry ground for new towns and cities on the river's bank. But the Mississippi wasn't, and still isn't, one to behave.

The rain didn't stop for months in the spring of 1927, and soon the saturated soil stopped absorbing the water. The resulting runoff inundated the Mississippi, and by April it was at historic levels. The levee system could no longer hold on. The first one broke on April 16 along the Illinois shore. Then, on April 21, the levee at Mound Landing in Mississippi gave way. Over the next few weeks, the entire levee system along the river collapsed, flooding 27,000 square miles, according to the Mississippi Department of Archives and History. Some places were submerged in 30 feet of water, and towns had to build elevated wooden paths for people to commute by foot. It took more than two months for the floodwater to subside.

By then, hundreds of people had died and nearly one million were left homeless. The flood caused more than $400 million in damage, or about $5 billion in today's dollars, and nearly bankrupted private flood insurers, who ultimately retreated from the market. Insuring property along our nation's coasts and low-lying areas was no longer a gamble worth taking, and many residents decided to cut their losses as well. Hundreds of thousands of people, mostly black, joined the great migration from the rural South to industrial cities like Chicago, New York, Detroit, Philadelphia, and Baltimore. The resulting clash of communities and cultures sparked a generations-long fight for equity and inclusivity that continues to this day.

Others, usually wealthy and white, remained and rebuilt. While private flood insurance became hard to find, most mortgages, including federally backed ones at the time, didn't require homeowners to have it. As a result, our rivers and coasts became largely unregulated landscapes for development, which were then gobbled up by homeowners who likely didn't know the risk. This continued for more than forty years, until 1968, when the federal government successfully established the National Flood Insurance Program, or NFIP, arguing they were best equipped to "balance the goals of mitigation and economic development in flood plains." But by then, engineered sandcastles already rose along our country's most precarious waterways, and the NFIP's flawed policy unintentionally incentivized more of this hazardous growth.

In order to attract customers and gain solvency, the NFIP had to make concessions, including subsidizing preexisting homes in the riskiest locations. The agency sectioned off communities into "zones," and every home in that zone—regardless of whether it was on the ocean or two miles inland—paid the same exact rate for insurance. The monthly bill a homeowner received in the mail each month was not based on the flood risk facing that specific property. It was the equivalent of a Wall Street banker and a fast-food worker being lumped into the same tax bracket.

Then there was the issue of what data was used to even determine risk. Flood zones do not consider future projections based on climate and habitat change. Instead, they are scaled based on historical data collected during a time when the impacts threatening our ecosystems hadn't even begun to hit the nation. In some cases, this data was more than forty years old, similar to New York City maps pre–Hurricane Sandy. "Maps do not forecast flooding. Maps only reflect past flooding conditions and are a snapshot in time. They do not represent all hazards and do not predict future conditions," Michael Grimm, then acting deputy associate administrator of FEMA's Federal Insurance and Mitigation Administration, told the *Washington Post*.

The NFIP's flawed policy made many properties appear safe when they were actually well within reach of water's grip. In fact, according to FEMA, "between 2015 and 2019, policyholders outside of high-risk areas filed more than 40% of all NFIP flood insurance claims." Officials later testified to Congress that these were properties in some areas FEMA had yet to map.

Something needed to be done to not only reduce a property owner's exposure (and FEMA's own financial risk) but also to deter future homebuyers from moving into the most threatened corners of the country, or at the very least make them aware of the risks in the likely event a future storm hit.

Following Hurricane Sandy, the NFIP—overseen today by FEMA—began developing a new method of assigning premiums for its 3.4 million flood insurance customers based on updated maps that analyzed each home's individual flood risk. "FEMA is using its capabilities and tools to address rating disparities by incorporating more flood risk variables. These include flood frequency, multiple flood types—river overflow, storm surge, coastal erosion and heavy rainfall—and distance to a water source along with property characteristics such as elevation and the cost to rebuild," the agency outlined on their website.

Under the so-called Risk Rating 2.0 plan, around 600,000 homes would see their annual costs fall, as the program restored balance.

But by and large, leveling the scales meant most homeowners would pay more, in some cases *thousands of dollars* more each year, over a yearslong integration period (federal law prohibits an increase of more than 18 percent annually). For flood-prone homes that had not been damaged by a storm, the breakdown worked out like this: For the majority of these homes, around 2.4 million in all, an increase in annual premiums would be capped at $120; 230,000 homes saw increases of up $240. Costs rose to $360 for 74,000 homes. And roughly 25,000 of the nation's most flood-prone homes would face additional annual costs up to $1,200.

There were discounts available. Homeowners who voluntarily stormproofed and elevated their properties could save hundreds and possibly thousands of dollars each year. Under RR2.0, however, these voluntary measures were mandatory for homeowners whose properties were significantly damaged or destroyed by a storm. According to FEMA, "all homes being substantially improved, or homes that have sustained substantial damage," as determined by the federal agency and enforced by local code and permitting, must be elevated or face punishing annual insurance rates as high as $20,000. And what if a homeowner, unable to afford this penalty, decided to just terminate their coverage and roll the dice? Not possible, at least for a majority of homeowners. Federally backed loans require every borrower living in a flood zone to have flood insurance. Seventy-five percent of all mortgages are federally backed, according to the U.S. Government Accountability Office, accounting for 33.4 million homes nationwide.

RR2.0 quickly won support from resilience experts, environmentalists, and the insurance industry, who said the program would help rein in risky building. But there was also bipartisan objection from some politicians who argued it would largely impact middle-class and lower-income enclaves—places like Breezy Point—where many couldn't afford a sudden spike in bills. Despite the protest, FEMA began rolling out RR2.0 in October 2021. It is a step in the right direction toward adaptation, but it's too early to tell if the new premiums will lead to safer development (and help restore these engineered ecosystems to their natural form) or turn America's coastlines into fortress-like playgrounds—climate change casinos where only the rich can afford the buy-in.

That's exactly what happened on a street called Harmes Way off a road named Sunken Meadow in Eastham, Cape Cod. The homes on this ocean spray–weathered strip of sand were once all owned by locals, until one day a small mid-century cottage was torn down and replaced by a new summer house built on stilts over the original home's foundation and topped off with a red roof. In the early 2000s, when construction was nearly done, I'd take a few of my father's beers from the refrigerator and drink my contraband on the unfinished home's deck overlooking Cape Cod Bay. Over time, many locals were squeezed out of Eastham as investors and buyers poured tens of thousands of dollars into raising their properties above the tempestuous whims of the sea. Living in increasingly unlivable places—engineering Mother Nature to man's will—is as American as apple pie, and now only reserved for those who have money.

But in a lesson first learned along the Mississippi and taught time and time again today, engineering and money can't always fight the rising tide. Conservative data collected by researchers at Britain's University of Bristol indicates major flood events will come 26 percent more often as sea levels rise at least one foot by 2050. And Risk Rating 2.0 is by no means foolproof. It will take years to update maps in some of the most remote areas like Leesville, and even then, these flood maps will still only measure risk based on historic data, not future storm projections.

Ultimately, it's a plan that sets up the next generation for disaster. Think about it. Future homeowners will likely opt out of flood insurance because historic data showed there wasn't a risk, only to be saddled with bills when the threat science predicted eventually hits. Nonetheless, a one-minute, twenty-three-second viral video from Rodanthe, North Carolina, puts our conundrum in stark relief. The clip shows a spring 2022 storm striking the historic coastline, lapping at a property as it hovers over the sandy beach on stilts. But the surging ocean appears to laugh, chopping those stilts down with ease.

According to updated records on Zillow.com, that house had a flood-risk warning of just 2 on a scale where 10 was the worst.

Fatal Attraction

"This might be the address. The cell number linked to it isn't accepting calls right now," my producer back in Los Angeles wrote in an email. The worst of Laura had passed, and now, finished with morning live-shot duties, my team and I were carving out an exit. It had taken dozens of men two hours to clear the path of fallen trees around our hotel, and their real work hadn't even begun. As Bill, my photographer, my audio technician, and I made our way south, we were stopped by roadblock after roadblock—everything from fallen trees to steel webs of downed power lines. Trees crushed cars and flipped over mobile homes. Roofs now covered streets, and traffic lights were snapped in half. We were headed to Holton Cemetery Road, where hours earlier Cindy had lived.

She was Laura's first confirmed fatality.

While my phone's GPS said it was an hour-long, 44-mile-drive away, I knew from traveling down similar roads in other hurricane-ravaged states that maps and ETAs only existed in civilized worlds. The delay was a welcomed one, and I secretly hoped we wouldn't be able to get through. While I've knocked on the doors of countless victims over the course of my career, asking a family to speak with me about a loved one they just lost never gets easier.

As Bill drove, I looked out the window and scrolled through my history of interviews for imperfections. I thought about my first television job out of college in Grand Junction, Colorado, and the doors I knocked on then. I revisited a brief stint in Milwaukee, Wisconsin, and my time in New York City and London. All was clear except for Miami.

I'd been there as a nightside reporter for WPLG, an ABC affiliate, and was assigned to cover a woman being released on bail for an alleged crime I can't recall. I do remember the crowd of journalists and photographers waiting for the door of the Miami-Dade detention center to swing open. Studying that door, I promised myself I wouldn't join the media crush should one happen, but when the woman finally emerged, and the cameras and reporters started chasing, I did, too. A dozen of us hollered out questions that went unanswered. This woman wasn't going to talk, but we had to give it our all so our bosses knew we at least tried. It was the TV-journalist equivalent of punching a time card. The woman was walking briskly, looking for someone who was supposed to come pick her up. Now, cornered against a fence and brick wall, questions getting louder and more rapid, she suddenly stopped, sat down on the floor, and began sobbing. She had survived jail, but not our firing squad. I looked around and could tell this was a first, even for the seasoned reporters in the pack. We all looked at each other, confused, then scrambled like a flock of seagulls after tearing apart a bag of potato chips on the beach.

The years since hadn't made me better at convincing people to talk. All that time instead taught me no interview is more important than someone's well-being. "When we get there, let's introduce ourselves, but if they don't want to talk, then I really don't want to push," I said to Bill. "Absolutely," he agreed as we got closer to the address.

Holton Cemetery Road

There is no street view of Holton Cemetery Road on Google Maps, but if you search for it on satellite, you'll get a bird's-eye view of a winding, narrow dirt path stitched through a blanket of old-growth trees. There's the cemetery the road was named after in a clearing on the left, and a pond where tadpoles turned into those throaty frogs across the street. At the dead end is the patch of swampy forest where the fireflies perform and, close by, the Miller home, streaked by the twisted shadows of trees. Google Earth said the view was captured on January 5, 2020. It was a digital snapshot of the Miller family mere months before it was broken forever.

The Holton Cemetery Road Bill and I now drove down was unrecognizable. Towering trees laid on either side, severed into four-foot-long pieces. It took the sheriff's department, firefighters, Cindy's father, and his neighbors five hours to clear the route we now drove down with ease. I remember it vividly. Parking the car. Telling photographer Michael Comfort to hold back until we introduced ourselves. Going to one house, only to be directed to the next one over. The one that looked undamaged from the front until you circled behind it.

The massive oak tree peeled the entire rear side, and I could now see into the spaces it once enclosed. On the right was the bedroom Cindy shared with her sisters. Through a layer of insulation sandwiched by Sheetrock was the bathroom, and through another layer was her parents' room, where they all rode out the storm together. "It's just an unbelievable tragedy. The poor child held on for as long as she could," said a man who identified himself as Richard Partridge, the family's pastor. "The paramedics and the coroner's office just left a few hours ago," he said. Cindy's death was fresh, and here we were, circling like vultures. I was trying to think of a graceful exit strategy. Should I express my condolences, hand him a business card, and just leave? "Did you help with the trees?" he asked after I waited too long to make my move. It wasn't the first time I'd been mistaken for help only to have to awkwardly reveal I was there to instead poke and prod with a camera recording it all. "We're actually journalists. We wanted to see if someone with the family would be open to sharing Cindy's story with us?" I told him. Richard looked deep in thought. "And we didn't realize just how recently this all happened," I said after a beat. "Her parents are too distraught to speak right now, but give me a moment," he said, pausing between each word.

Bill and I stayed where we were, frozen in front of the cross section of someone's life. The book Cindy was reading in her final moments was on the grass, partially dusted with insulation. A mattress that saved James's life by absorbing his body when part of the oak fell on him was now leaning against the house. But his daughter did not have an airbag that night. The remnants of the tree that killed Cindy were stacked in pieces close by. "I'd like you to meet someone," Pastor Richard said, returning with a young girl. "This is Cindy's older sister, Nellie. We talked, and I think it would be helpful for her to share a few stories about her sister." Nellie's younger sister Maddie also joined by her side. She was still wearing the same clothes she had on when Laura barreled down, including a black Batman sweatshirt. In her hands was a framed picture of Cindy she managed to pull from the wreckage. They looked identical, down to similar framed glasses.

"I tried to wake her up, but she wouldn't wake up. She was pinned under the tree," she said. Nellie told me about Cindy's dreams of going to Harvard and her desire to become a microbiologist. She told me how much her family loved each other and how she'd bicker with her sisters but always quickly make up. She told me about the fireflies and Cindy's love for animals and her passion for reading. As Nellie spoke, she mixed-up tenses. The past had yet to catch up with the present, but it was all beginning to sink in. "She was special. She was going to do something really big."

Lessons Learned

According to NOAA, the top seven most destructive hurricanes in U.S. history, even when damage estimates are adjusted for inflation, all occurred in the past twenty years. Michael, Wilma, Ivan, Ian, Ike, Irma, Sandy, Maria, Harvey, and Katrina collectively cost more than $800 billion in damage. The 2017 trio of Harvey, Maria, and Irma had a combined cost of $268 billion, or 31 percent of all damage since 1980.

This price tag reflects two conflicting trends: Hurricanes are intensifying as predicted by scientists; and more people are developing and moving into these flood-prone zones. In the last four decades, the population of counties along the Gulf and East Coast increased by about 160 people per square mile, compared to 26 people per square mile in nearby inland regions, according to risk mitigation firm Aon. The population of upstate New York, by comparison, which is considered one of the country's regions most insulated from ecosystem collapse, saw either flat or declining growth. Meanwhile South Florida—threatened by sea level rise and already experiencing extreme flooding, heat, and megahurricanes—saw a 10 percent population increase between 2014 and 2018. Nationwide, an estimated 41 million Americans live in areas designated as flood prone, according to a 2018 study coauthored by the U.S. Environmental Protection Agency (EPA)—the first and only study of its kind.

When shit hits the fan in these threatened ecosystems, as it inevitably will, FEMA steps in to help, providing 75 percent of the funds needed to rebuild *public* infrastructure. That money comes from taxpayers. Similar to FEMA's flawed flood insurance policy, families in some of America's safest places, like upstate New York, subsidize those living in the nation's most threatened, to the tune of hundreds of dollars per household, per tax season—and it's growing. It's like that friend who orders the most expensive thing on the menu—with a side of two martinis and a glass of port—and then suggests splitting the bill. The federal government has caught on and pulled back slightly on this gravy train. Today there's more steps on the road to applying for federal grants and more scrutiny when approving them, and that has incentivized local municipalities to reduce their risk. Some, like Florida's Babcock Ranch, have been incredibly successful.

Babcock Ranch is an 18,000-acre master-planned community built on old cattle land outside Fort Myers that first opened its doors in January 2018, quickly attracting a population of around 5,000 people, many of them escaping coastal lives threatened each year by hurricanes. Everything from homes and business to streetlights and park benches was designed and built to withstand a Category 3 hurricane packing 145-mile-an-hour winds. The land was even elevated 25 feet above sea level, far out of reach of coastal storm surge, and the developer used natural landscaping to help prevent flooding and erosion. Developers also partnered with Florida Power and Light to make Babcock the first entirely solar-powered city in the country, along with placing all of its utility lines underground. When Hurricane Ian made landfall as a Category 4 storm in 2022, it leveled entire coastal communities nearby, but Babcock, which took a direct hit, didn't even lose power. But of course, it's easier to fortify an ecosystem when you're starting from scratch.

The Florida Keys are an archipelago of forty-four limestone islands strung together by a thin 113-mile-long thread known as scenic Highway 1. Limestone is a porous rock and perhaps not the best foundation for humans to anchor their lives to, but in the early 1800s, when the first homes barnacled this southern slice of sedimentary Swiss cheese, the impacts from industrialization had yet to materialize. Today, the Key's kitschy mix of tourist traps, seafood shacks, famed scuba diving, and deep-sea fishing attract millions of visitors each year and even once hooked writer Ernest Hemingway, who took up residence along with a colony of cats at the end of the line in Key West. By the time rising seas began seeping through and flooding out life above land with regularity—around the turn of this century—it was too late. More than 70,000 people call this precarious paradise home, according to the U.S. Census. And with an average median household income of just over $72,000, costly retrofitting isn't an option for many. Today, 79 percent of all homes in Monroe County are vulnerable to flooding, and by mid-century reaching those still standing will be impossible . . . and I don't say this in a "scientists predict" kind of way. It really will be impossible. According to a 2022 roads vulnerability study commissioned by county officials, 49 percent of the publicly owned roadways will be subject to sea level rise by 2045. Monroe County officials have been calculating how much it will cost to raise an estimated 300 miles of highway, main streets, and back roads. A sliver of road on a Key called Sugarloaf shows the many hurdles ahead. The county determined that to elevate just a three-mile section, one foot would cost $75 million and only keep the path dry until 2030. To protect that short stretch against 2060 sea level–estimates would cost taxpayers *$181 million*. And what happens if a major hurricane damages or destroys it?

Local officials said it could take years to get funding from FEMA, and even then the community would still be on the line for 25 percent of the project's total cost, which it can't afford. Even seasoned politicians could not find the right words to talk around the alarming truth. Many parts of this American Paradise could eventually be cut off. "We have to work together. Government can't solve everybody's problems. We can certainly help, but it's only with the businesses and the residents working with us and helping out that we're going to have really true resilience in the Keys," said Rhonda Haag, the county's chief resilience officer. In a board meeting, one county official suggested it might be most cost-effective to offer impacted residents a ferry, water taxi, or some other kind of aquatic transportation. But as basic infrastructure disappears—making it harder for residents to literally go about their business—and as the threat of stronger storms looms, there's growing concern entire towns in the Keys will vanish, too. It's happened before.

Asleep at the Wheel

Fair Bluff, North Carolina, weathered its fair share of storms, beginning with an economic one in the 1970s when the local tobacco plant shut down as the nation's rate of smokers began its steep late-century decline. Fair Bluff reinvented itself a few years later, becoming one of the leading suppliers of vinyl flooring in the state. But most people were so focused on keeping Main Street open that they paid little attention to the occasional flooding that became more routine over the years. The NFIP's outdated flood maps didn't even include the area. Even the town itself chose not to get flood insurance for its built facilities. "I try not to point fingers, because if I did, I'd have to start with myself," town planner Al Leonard told me. "We all sensed something was up with flooding. We fell asleep at the wheel." Well, 2016's Hurricane Matthew was the alarm clock. The deluge submerged the entire town in chest-high water as the nearby Lumber River crested over its banks. "The cushions from chairs and couches were floating around," Al recalled. "Every document was completely destroyed. We didn't own a ballpoint pen or a paper clip. The mayor's truck was our new town hall." The real town hall, along with the police department, firehouse, and a quarter of all homes in the city, were destroyed. As town planner, it was Al's job to rebuild . . . from scratch. "The first year or two it was almost like you were in combat. You're trying to survive, wake up the next day alive and not dead.

We were so tied up with trying not to die as a municipality that we didn't think of much else."

The blueprint for a new path forward hadn't even been finalized yet when Hurricane Florence hit two years later, following a path of destruction nearly identical to Matthew's. Physically, there was little left of Fair Bluff to damage, but mentally, Florence was the final straw. "People just said enough is enough," Al told me. "When two storms hit so close together, you just know another one will likely come in your lifetime." Seemingly overnight, the close-knit community began to unravel as people who had been committed to rebuilding the town together instead opted to take federal buyouts and move to safer ground. "The first thing that we saw when our people left was the town's only grocery store go out of business. The owner of the store said there weren't enough people anymore to support it. The drugstore did not reopen. The hardware store did not open," Al recounted.

As of 2022, there were only around 500 residents remaining in Fair Bluff, down from 1,000 before. Of the forty buildings that lined downtown, only one was occupied: the post office. Everything else was shuttered. Stores didn't just lose customers; the town lost its ability to fund basic services as the tax base disintegrated. "Obviously when the government buys a home and takes it off the tax books, that's revenue we'll never see again. And as a small town, that means one less police officer or firefighter," said Al. Unlike Paradise, California, which received billions of dollars from PG&E for the utility's role in starting the Camp Fire, there was no single person or company to blame for Fair Bluff's demise.

Fair Bluff is a case study in the challenges the federal government will face as it responds to environmental disasters. Buying out communities may sound equitable on paper—after all, the residents of a town don't have much say over environmental policy on the local level, let alone power on the national and global stage to effect change at home. But the program is far from a solution on its own. From a financial standpoint, it's expensive for the federal government to buy even small towns like Fair Bluff. From a sociological perspective, the loss of identity, history, and community can be paralyzing, especially when people are relocated to a new town with different customs and traditions. For those reasons, FEMA has only relocated a few communities in the agency's history, but it may not have an option in the future as more towns suffer from collapse. President Biden's $1 trillion infrastructure bill provided the agency's flood relief program with $700 million taxpayer dollars to help buy homes, neighborhoods, and entire towns at risk of climate catastrophe. As more people knowingly and unknowingly move into at-risk regions, it's likely this program and similar ones will expand.

In Fair Bluff, as Al told me, the current plan is to apply for federal aid to buy all forty buildings downtown, then demolish them and build green space. Fair Bluff would then relocate the businesses to higher ground that didn't flood in the two storms. The new location will be appropriately called "uptown." But for now it's just a dream that will require tens of millions of dollars in federal funding, which Al says is not guaranteed and is more difficult to argue for when a community is thinning out. "The signs for concern have always been there. We're right on the river, which has flooded before, but obviously not like this. We'll never turn our backs on her again."

Increasingly, America's small riverside and coastal towns have become beacons, illuminating our nation's vanishing edges and the challenges that come with moving to safer ground. More and more communities will face similar struggles in the years to come. Where water meets land, there is good reason to be concerned.

A Final Resting Place

It didn't take long for the Spanish moss, duckweed, salvinia, and spider lilies to begin covering the wounds created by Hurricane Laura. Where the sun touched, they grew, and on this day those rays of light flickered through the gently swaying cypress and oaks overhead as Mary, James, and their daughters made a thousand-foot journey through their bandaged neighborhood to visit the place's newest landmark. Loose bits of gravel and dirt crunched under their feet as they walked, solemnly, to the cemetery that gave Holton Cemetery Road its name. Here, among headstones with dates that told stories of long lives lived well, was one about a dream cut short. "Cynthia D. Miller Mar 10, 2006–Aug 27, 2020."

Reverend Richard Partridge officiated the intimate graveside service. James stared off into the distance. His eyes were red and glazed over, the look of hindsight and the grief that follows when playing things back in your head still doesn't change the outcome. As Mary and James tried to find peace, they were especially grateful for what they had discovered among Cindy's belongings earlier in the week . . . a message, they believed, from the other side. In what was left of Cindy's storm-scattered room, James recovered a corkboard where she had used thumbtacks to pin and stick Post-it notes with quotes from her favorite books, movies, and musicians. Among Bowie, Gatsby, and Percy Jackson was one from World War II singer Vera Lynn. "We'll meet again. Don't know where, Don't know when. But I know we'll meet again some sunny day."

Cindy's family and friends blanketed her gravesite with flowers and photos during the hour-long service and made silent promises to visit again soon. Later that afternoon, as the sun began to set, the crickets and frogs started their own serenade, and as the sky faded to a deep cerulean, even a few fireflies lit their torches.

PART THREE

AIR

Foote's Notes

Not a single confirmed image of Eunice Foote exists, and the memories of her have mostly slipped through the holes that often riddle oral histories. The few stale crumbs that have survived all these years paint Eunice's blank canvas with features that resemble other female trailblazers that came after her—people like Nellie Bly, Freya Stark, and Amelia Earhart. In 1856—around the same time prominent American doctor and professor Charles Meigs wrote a woman "has a head almost too small for intellect but just big enough for love"—Eunice broke through glass ceilings to challenge man's understanding of both women and Earth's atmosphere. And then, she seemingly vanished into thin air.

Eunice was born in Goshen, Connecticut, on July 17, 1819, and raised on a farm with her ten brothers and sisters. As a teenager, when most girls were learning homekeeping, social graces, and etiquette, Eunice attended Troy Female Seminary and spent much of her time experimenting in its small chemistry lab, a place women generally were not welcomed outside the protective walls of an all-girls school. She was so committed to her studies, she even audited university chemistry classes and sat in on lectures. Her passion for science, and her drive to shatter barriers that kept women like her out of laboratories, continued into adulthood.

In 1841, she married her husband, Elisha, an inventor and fellow amateur scientist who shared her passion for Earth studies and commitment to women's rights. The couple became prominent feminists, playing key roles in the 1848 Seneca Falls Convention that launched the American suffrage movement. Eunice and Elisha's signatures, one of the rare crumbs that exist today, can be found beneath the convention's manifesto, which they helped edit. Remarkably, a man would later get all the credit for Eunice's greatest discovery: climate change.

In the early 1850s, Eunice carried out a simple experiment in her kitchen. She filled three glass cylinders. One with moist air, another with dry air, and the third with carbon dioxide. According to her notes, she placed a thermometer in each container and left them in the sunlight. What she observed was groundbreaking. The cylinder with moist air was hotter than the one with dry air. The cylinder with carbon dioxide was the hottest of them all, and took the longest to cool down once removed from the light. "An atmosphere of that gas would give to our earth a high temperature," she concluded in her 1856 paper "Circumstances Affecting the Heat of the Sun's Rays."

Eunice's prophetic discovery came just as the Western World had been transformed by a hundred-year industrial revolution. Few at the time questioned, or were incentivized to worry about, the impact carbon dioxide and other gasses released from factories would have on Earth's atmosphere. And in the end, that may be why her paper, even with the endorsement of Joseph Henry, a respected researcher with the Smithsonian Institution, received little attention. Joseph later told the *New York Daily Tribune* that, while the experiments were "interesting and valuable," the scientific community struggled "to interpret their significance." Being a visionary is lonely when you're the only one who can see. Eunice published one more paper, on the electrical energy of gases, before abandoning science altogether.

While she was the first to observe CO_2's impact on rising temperatures, a scientist in England named John Tyndall got all the credit when, three years later, he published his now famous paper "Note on the Transmission of Radiant Heat Through Gaseous Bodies." It was a more muscled body of work than Eunice's, but both shared identical bones. Published in the British journal *Proceedings of the Royal Society of London*, John's research placed carbon dioxide along with methane, water vapor, ozone, and nitrous oxide in a notorious group that today we call "greenhouse gasses." Like the glass panes of a greenhouse, every single minuscule molecule in his group of gasses acted as a shield, preventing heat from escaping into space. With time, the built-up heat would evaporate all the moisture, creating an endless cycle by pushing more heat-trapping vapor into the air and wreaking havoc on the soil while simultaneously baking the delicate cells of any life therein. In a letter he wrote to the Royal Society while submitting his research, John emphasized "Nothing, so far as I'm aware, has been published on the transmission of radiant heat through gaseous bodies." John's findings, later described as the "greenhouse effect," were celebrated and soon universally adopted by the scientific community.

Today Tyndall's name adorns everything from mountains in California and Australia to glaciers in Colorado and Chile . . . even a crater on the moon. As for Eunice, you'll only find her name etched on a stone mausoleum at plot 8379 section 34 in the Green-Wood Cemetery in Brooklyn, New York. She has rested there in peace since September 30, 1888.

So, did John flat-out steal Eunice's research, or was this mere coincidence? Communication at the time was limited to only letters and telegraphs, and American and European scientific communities in the 1800s were largely isolated islands. But several historians point to a curious sequence of events that appear to trace John directly to Eunice. Around the time of his discovery, John was an editor at *Philosophical Magazine*, a British publication, and responsible for the content printed in each issue. In 1857, the magazine republished research on meteorology that was first printed in the November 1856 issue of the *American Journal of Science and Arts*. The author was Elisha Foote, Eunice's husband. It was the same issue of the same journal in which Eunice's paper on climate change appeared. The respective research was printed back-to-back.

Regardless of what you believe, coincidence or straight-up theft, neither John nor Eunice's experiments forecasted how fast CO_2 emissions would build up and transform what was at that point a stable atmosphere into a chaotic and unpredictable one. Up until the twentieth century, our planet's climate had changed at a relatively slow and steady pace for millions of years—so slowly that life on Earth was able to adapt and even thrive. But add planes and trains and automobiles and boats and cell phones and computers and televisions and light bulbs and washers and dryers and whatever else you rely on that uses gas and electricity to the equation, and you've got the equivalent of 6,000 years of greenhouse gas (per Eunice's and John's presumptions) in just the past six decades. According to NOAA, since 1960, concentrations of CO_2 in our atmosphere have tripled to levels not seen since before humans roamed Earth, when our oceans were up to 82 feet higher. While today's cloud of heat-trapping carbon and other greenhouse gasses hangs over the entire globe, binding disparate communities to the same shared fate, the invisible glass panes are impacting different ecosystems around the world in different ways. Mayfield, Kentucky, and Red Lodge, Montana, however, do have one thing in common. Most didn't see what hit them coming. What Eunice Foote first observed in her controlled kitchen experiments 150 years ago, and what John Tyndall later elaborated on in his own research, is now unfolding with alarming accuracy and speed in the natural world.

Parallel Lives

The singer looked promising as he prepared for his set when I entered The Local, the college honky-tonk where the woman at the front desk of the Nashville Marriott recommended I grab a bite. "It's past Centennial Park near the Family Barber Shop. They've got great wings and live music," she said. And so I headed there, ordered a basket with blue cheese and extra hot sauce on the side, grabbed a draft IPA, and sat at one of the high tops. The singer was wearing a pair of jeans, worn-in boots, and a Patagonia vest over a red flannel shirt. A baseball cap covered his eyes. It was an understated, more nuanced outfit compared to the cowboy costume the performer before him was wearing. My guy—the one I was now rooting for as I imagined a talent scout listening in the audience—looked like he'd just gotten back from a fly-fishing trip. He opened with a grass-is-greener allegory by Travis Tritt about a son wanting to leave the family farm for the big city. I wrote down the lyrics in the notes app of my phone to revisit later: "I think it kinda hurt him when I said, 'Daddy there's a lot that I don't know. But don't you ever dream about a life where corn don't grow?'"

He only sang covers, but with the tone of ownership. Not bad. He was on his fifth song when I started to make my way back to the hotel. It was getting late, but I had only arrived in Nashville a few hours earlier and was still on West Coast time. The thought of going straight to my carbon copy corporate hotel room and watching reruns of *Shark Tank* on TV wasn't the nightcap I was looking for, so I passed the elevator and headed to the warm-wooded, backlit-bottled bar to better acclimate to my temporary base camp. There, a man in a suit ordered a drink from a jewel of a bottle. "It's a rhombicuboctahedron," he said after noticing me stare for a few seconds too long. "That's the shape of the bottle," he politely volunteered when the vague response that I usually deploy when I'm clueless about something that could be common knowledge—"oh"—slipped past my lips. It looked like a crystal grenade with amber liquid splashing inside and a cork stopper with a metal horse and jockey on top. The bartender then placed one of those massive you're-about-to-pay-a-fuckton-for-this-drink kind of square ice cubes in a plaid-etched glass and, handling the rhombicuboctahedron like it really was a grenade, slowly eyed her pour. The precious elixir sparkled as it waterfalled down the glacial slab. She stopped when the ice was submerged about a quarter of the way. "I'll have what he's having," I said when it was my turn, not even knowing what it was he was having. "Good choice," the man in the suit said. "Blanton's is some of the best bourbon you'll find and it's rare to get it in a bar. It's probably even harder to find now with all the damage in Kentucky."

That damage was what brought me to this twangy corner of the country. On December 10, 2021, unseasonably warm air sparked an outbreak of tornados that killed dozens of people and mowed historic towns down to their roots. Churches and theaters and restaurants and cafes were stacked in piles of brick and wood and twisted metal. Those who lost loved ones were visiting fresh memorials. Those who lost their homes were picking through the rubble for anything that survived. The people whose homes were still standing, the "lucky ones," who were told they didn't qualify for temporary federal housing, roamed their broken communities like Bambis.

The following morning, I'd be meeting my producer in the hotel lobby and then heading out to report on the cleanup and recovery effort. Mayfield, Kentucky, was only a two-hour drive away, but the monster tornado that had ripped through town rewound it by centuries. Soon I'd be leaving behind civilization for something . . . else. For now, though, I was drifting through the waiting room of the in-between, a state of existence before launching to a place most everyone else has fled. It's an acclimation that requires adapting from the social norms of healthy ecosystems to the lawlessness, destruction, and pain found in cities and towns that have burst. Here, in this bar on this stool, I was cautious not to open my big mouth, aware that those who have never seen total destruction have a morbid fascination that people on the flip side of luck don't exactly appreciate. In fact, I usually avoid the kind of eye contact that compels a suited stranger to make the kind of small talk that inevitably warms up to the kind of questions that, when answered, attract a small crowd like a cheap circus act. Step right up, folks!

I've gotten good at blending into the geometric patterns of the rugs that seem to line every one of these hotels. I first earned my squares and circles and weird swirly things years earlier while grabbing breakfast with a different producer at a different hotel bar. We were covering a wildfire and mapping out the areas that had already burned when a man in his thirties yelled at us from the next table. "Some of us have lost our homes. Have some fucking respect." That day I graduated from a croaking frog to a quiet chameleon.

There's also a fair amount of guilt that surfaces while spending time in the in-between. To be sitting in a fully stocked bar with carefully sculpted frozen water talking about ridiculously shaped bottles of liquor when the people I'd soon be talking to didn't even have running tap—was numbing. The bill from my drink made me feel again. A single shot was $39. I said good night to the bartender and suited man and took my glass to my room, where I savored every sip, right down to the last drop of that melted cube. It really was some of the best bourbon I'd ever had. As I settled in for the night, I googled the lyrics to that Travis Tritt song.

The song rung true in more ways than one. It's becoming increasingly harder to find greener grass in our changing climate, and a growing number of people, no matter how far apart they are on the map, are living eerily similar, albeit parallel lives connected by the air we all share.

Fork in the River

It was seven p.m. on June 12, 2022, and the rain was a welcomed break in a weekend wave of unseasonably high temperatures for Montana. Just one day earlier, the sidewalks on Broadway radiated waves of heat that distorted the feet of tourists as they walked in and out of the small-town specialty shops, such as the Montana Candy Emporium, Lewis and Bark's Outpost, and Grizzly Peak Outdoors. The American flags that were mounted on every lamppost in preparation for the Fourth of July parade were now flapping violently in the heavy wind like sails in a squall. Montana's famous blue sky was soon brushed shades of licorice, and the streets, whitewashed from years of winter salting, were Pollocked back to their original dark gray by drops of rain. The whole town of Red Lodge smelled of cooled concrete.

Fire chief Tom Kuntz was in his Jeep pickup with his two daughters when he got the call. There was a gas line break reported near the Rocky Fork Inn on South Broadway. "I'll head over now," he told the dispatcher on his radio. As they got closer, he noticed water was beginning to stream down the road. There was no way the source of it was the sky, he thought. The rain was strong, but not *that* strong. By the time he reached the intersection of South Broadway and 19th, he discovered the origin. Rock Creek, normally the kind of place kids looked for trout in the shallow trickle of pebbled water, had swelled to the top of its 12-foot-high steel and concrete river wall—the one that had been built less than thirty years ago to last more than a hundred— and was spilling into the street like a bathtub left running too long.

As Tom inspected the scene, he realized the gas line had broken along with the partially collapsed wall it had once run alongside. Tom couldn't believe his eyes. In even the worst storms, the creek never rose more than a few feet. "What caused such a catastrophic surge, and how could water, no matter how much there was of it, tear apart a two-foot-thick wall?" he wondered. Rock Creek expanded with each passing minute, and by ten p.m., what was left of the wall and streets and sidewalks and business entrances disappeared under a frothy swirl of chocolate milk–colored water. Soon Broadway, a northern gateway to Yellowstone National Park, was a Rocky Mountain Atlantis. Tom watched the water level rise, knowing only Mother Nature was in control of when it would stop.

Eventually the fierce current carved its own path down Broadway before flowing right onto 15th Street and into homes on Platt, Haggin, and Daly Avenues. The water seeped through windows and doorframes at first, turning wood floors slick and carpeted ones soggy. Eight-year-old Bri Beekman was playing in her basement bedroom when her cat suddenly jumped with a loud "scream." By the time residents processed what their eyes were seeing, the white-water rapids that had taken over Broadway swept through their homes, shattering windows and even tearing off entire walls. The power flickered out sometime before midnight, and the only light inside living rooms, dining rooms, bedrooms, and kitchens came from flashlights that refracted off the torrent and cast waves on ceilings and walls. The rush of water also sounded like thunder. No, that wasn't it. This was something different. Tom also noticed it on Broadway. The sound of rocks bashed together under crashing waves. But as Tom continued to assess the damage, he'd learn the reality defied belief. These weren't just "rocks" tumbling, swept up in a current in the middle of his town. They were 1,000-pound boulders—probably millions of them—hammering away at Red Lodge. As families escaped to higher ground, they were left to wonder: Where was all this water coming from?

Stirring Up Monsters

Walker Dawson has the kind of name that sounds exactly as he looks. He's a tall and handsome J.Crew–type whose voice softens when his curiosity is piqued. "So tell me, who exactly are we meeting today?" he said in a whisper as we loaded our gear in his car outside the Nashville Marriott at seven a.m. I could see in his eyes he was already trying to plan out the shots we would need for our story. As a digital journalist, or DJ, Walker wore many hats. On this trip he'd be both my producer and photographer. It's a job not many could do well, but Walker isn't like many. He once spent a summer with his father, a documentarian, visiting libraries in all fifty states. He's smart, and loves the little things others usually overlook or don't appreciate. He also likes to listen to obscure podcasts I've never heard of . . . something that would come in handy as we left behind our Tennessee base camp for the road ahead. For the next week, we'd be following the path of destruction left behind by one of the nation's largest tornados. It had been three weeks, and the active search and rescue for victims trapped in the rubble had transitioned to recovery. Ninety people were killed, many crushed under the weight of pancaked homes and businesses, like the candle factory in the town of Mayfield. Thousands more were homeless. We'd spend time with FEMA as they cleaned up and helped survivors, but the rest of our reporting, for the most part, depended on where the wind took us.

That wind, or at least the aftermath of it, could be traced in the trees. Hundreds of thousands of them were bent, limbs broken, in the direction the storm passed. "Had it moved just slightly east or west, the damage wouldn't have been nearly as bad," Walker said. "Those places are much less populated." That's exactly what meteorologists thought as they watched the air heat up and the storm form in the early hours of December 10.

———————

The single-story beige brick building off Old Highway 60 near the Barkley Regional Airport in West Paducah, Kentucky, looked more like a rest stop than a place housing lifesaving technology. It was five a.m., and a lone outdoor lamp cast a cone-shaped streak of yellow over the main entrance. The lights inside had been turned off for the night, but the lab glowed as computers processed radar data and flashed the pixelated results over maps displayed on a mosaic of television screens. In this predawn hour, the computer monitors in the National Weather Service's West Paducah field station flickered with patches of comforting blue. Just a few light mists. The map of Kentucky and surrounding states, updating every few seconds, was clear of any significant disturbances. But soon, the volatile hot air of a storm system forecasters warned about for a week entered West Peducah's range, turning the soothing cerulean edges of the digital bird's-eye view the ominous red of heavy rain and thunderstorms.

Meteorologist Pat Spoden had seen a lot in his thirty-four years covering the South's increasingly warmer, wetter, and unpredictable weather, and his entire team of nine was ready for this moment. They arrived at the field station around 7:30 a.m., brewed the first of what would be many pots of coffee that day, and fixed their eyes on the wall of monitors around them. It was their job to analyze the Doppler and issue public advisories based on what they saw. Thunderstorms don't typically require an all-hands-on-deck response, but add in an unusual winter heat wave and you have a recipe for disaster. Temperatures were going to reach the 70s. Pat had a pit in his stomach all morning. He knew conditions weren't good. Tornadoes require a few key ingredients to form: warm, moist air in the lower atmosphere; cold, dry air in the upper atmosphere; and wind to twist all that air up and set it into motion. The thunderstorm they were watching was the kind that could stir up monsters.

Pat stroked the peach fuzz on his head as he quietly scanned the radar images like a concerned oncologist reviewing a CAT scan. The cancerous crimson on screen reflected off his thin metal-framed glasses, casting a bloodshot glow around his eyes. By 10:19 a.m., lead forecaster Chris Noles knew there was no more time to wait. He fired off a notification to the National Weather Service's X account.

"This could be a significant severe event with a strong tornado or two across our region. Think about what you would do now. Better to err on the safe side."

A similar, more succinct message was beamed to phones, scrolled under morning television talk shows, and read on FM and AM radios. It was still unclear if all the ingredients would mix together, and if so where they would track, but one thing was now certain: This storm wasn't just going to blow over. In the hours that followed, disorganized patches of red pixels over parts of Arkansas started to consolidate. The storm meteorologists had predicted, and the kind of tornado they worried it would birth, was beginning to take shape. In *this* ecosystem, the combination of warm and cold air and wind was about to cause an explosion in the sky.

Delirium Nocturnum

I remember the call. And I remember ignoring it, too. It was 3:30 in the morning and my husband, Ivan, and I had only gone to bed two hours earlier, after having a few friends over for a dinner that drifted into late-night experimentation with brightly colored liquors. My head felt like a garbage disposal and my eyes were sealed shut. Was the call real, or just part of a dream? Maybe the dark, static hours when normal people sleep were playing tricks on my dehydrated brain.

"Amor, your phone is ringing," Ivan said, confirming I was, in fact, not imagining things.

"If it's important they'll call back," I mumbled as it went to voicemail, hoping a friend in London or New York had forgotten what hour it was in Los Angeles. It wouldn't have been the first time. Without skipping a beat, the phone began ringing again. Whoever was on the other end was *fucking persistent*, I thought as I fumbled for the phone. The glowing screen revealed it was my bureau chief, Joelle Martinez. I knew my dawn had arrived early.

"Hello," I whimpered into the receiver; half my face was still sunk into my mattress.

"I'm sorry to call so early," she said sincerely. "There's major flooding in Yellowstone National Park and there's evacuations in nearby towns. It looks really bad. Terri wants us to look at flight options that get you in before the *Evening News*." A quick scroll through social media showed wild rivers streaming down mountains in and around the national park and into towns around the northern border. Entire homes had been swept away in the currents and mudslides. Others seemed to explode when hit by the rocky waves. "How could rain cause so much flooding, and how the hell was I going to get there?" I thought as I looked at the time and considered the journey ahead.

It's surviving the nocturnal delirium that follows such calls—not the live shots, deadlines, or politics of network news—that ultimately determines if a person is cut out for this kind of job. If you don't know what I mean, try waking up at 3:30 in the morning, booking a last-minute flight while brushing your damn teeth in the shower (multitasking is key), securing a rental car and hotel, packing a bag in the dark with enough clothes to last a week while struggling to find any clean socks (admittedly, I've gotten lazier about keeping my go-bag in "go" shape), and ordering an Uber to LAX to catch the first flight out . . . all in less than thirty minutes and before a single cup of coffee. I once heard a story about a correspondent who told their bureau chief to go to hell and then turned off their phone. Such decisions made in the delirium nocturnum are never wise, and that correspondent didn't get many calls after that. Cursing, I find, sedates my inner saboteur. It also helps to have an amazing husband—to say Ivan is a saint for always insisting I turn on the lights to see what I'm doing and reminding me that I love what I do as I release my evil spirits is an understatement. The truth is, I wouldn't be able to do any of it without him. (Yes, I'll be ripping out, framing, and conveniently referring to this preserved page often as proof I don't take him for granted.)

And the other truth is, as much as I curse my vertical existence in those horizontal hours—and I do curse a ton—I wouldn't trade it for the world. Usually, I remember that right around the time whatever *fucking* plane I'm on takes off for whatever beautiful city I'll be parachuting into. "I love you" I texted Ivan like I always do once I landed in Bozeman. My brain still felt shriveled, and my eyes were red, but I managed to nap thirty minutes on the flight and secured a cup of coffee while I waited for my rental car.

"We only have a Toyota 4Runner. Will that work for you?" asked the woman at the Avis counter.

I took a few beats to carefully assess the repercussions of my response. The cogs in my head were still spinning slowly. "Does it have four-wheel drive?"

"It's not called a 4Runner for nothing, hun."

"Oh . . . right," I said with a smile. I got into the car and plugged the town where I'd be meeting my producers into the GPS. Red Lodge, Montana.

The Beast

Pat Spoden had been worried about the forecast for a week leading up to that fateful Friday, December 10. The seven-day outlook
roller-coastered between winter and summer in a matter of days.
On Tuesday in nearby Searcy, Arkansas, the weather was partly
cloudy with a high of 48 degrees. Typical for that time of year. But
by December 10 the air's temperature spiked to a record 79. That
afternoon, a window at the White County Courthouse was cracked
open and a rickety GE desk fan oscillated the air that smelled of
Christmas wreath. The red-and-white-brick landmark building was
strung with lights for a holiday that felt seasons away. The hot air
brewed a thunderstorm around three p.m., and those who looked
up remembered the moment a ghoulish-green cloud blemished the
paling sky. There was also an unusually sweet and pungent smell
in the air: the aroma of lightning violently splitting nitrogen and
oxygen molecules.

The Doppler radar indicated the thunderstorm forming over Searcy
could grow into a supercell with the power to produce not one, but
multiple strong tornadoes. Shortly after three p.m., after several hours'
warning to the public to be prepared for anything, meteorologists in
the West Paducah field office escalated their guidance and issued their
first tornado watch. Pat and his team had already identified what they
believed was a small, short-lived twister, and there was good reason to

think other, larger ones would come as the weather system continued
to blend on high.

The wind was now gusting to over 75 mph. Heavy rain became golf
ball–sized hail before the darkening thunderclouds, growing heavier by
the minute, began draining their pent-up energy in a slow downward
spiral. One person described what they saw as a drill poking through
the heavens. The young tornado kept boring through the raw electrified
air until it reached the bone of Earth, spitting out soil like sawdust as
it commenced its forward march. Searcy was spared by its wobbly first
steps, but the tornado gained more energy as it carved out an interstate,
first eviscerating a nursing home in Monette, Arkansas. It then crossed
over the Mississippi River, shaved the western edge of Tennessee,
and bulldozed through Kentucky. It was dark now and the tornado
vanished under night's veil, but every few seconds a flash of lightning
revealed the monster in this haunted house. A truck driver captured
Mother Nature's pulse on his cell phone along with the anxiety that
possessed each moment of darkness when it was anyone's guess where
the tornado would be when the lights briefly came back on.

Pat compared the images that strobed across his radar screen to the
doomsday scenarios he'd only seen in computer simulations during
trainings. The storm had supersized and generated several twisters, but
the monster that formed just outside Searcy was the most worrisome.
It had quickly grown, creating a vortex of wind that reached speeds
of 190 mph. According to the Enhanced Fujita Scale, named after
Dr. Tetsuya Fujita, who developed it to measure a storm system's
strength, the tornado was just 10 mph shy of an EF5, the highest
possible ranking. But Pat was most concerned about its size. Radar
was showing the twister was at least a mile wide, and there was no
indication it would dissipate. Nothing was going to stop this thing.
Pat also noticed a troubling patch of deep purple pixels forming in
the center of the twisted red like knuckle marks in a deep bruise. It
was the color of high-density debris sucked up by the storm's vortex.
Cars, mobile homes, siding, roofs, street signs, Sheetrock, window

frames, and animals were spiraling as high as 30,000 feet in the sky, the elevation most airlines fly.

The average tornado lasts less than ten minutes and travels about three miles before vanishing back into the air that made it, according to NOAA, but this one had already torn through several counties and there was plenty of atmospheric fuel to keep it going. Shortly before eight p.m., Pat's colleague Chris alerted the public. "You need to be ready to invoke your Shelter in Place plan, as an intense Tornadic storm may arrive in the 8:45 to 9:45 p.m. time frames." Twelve minutes later the message got more urgent. "Tornado still on the ground. This storm is headed for west KY. Plan now!" Pat and his team watched as the tornado, which they now called "The Beast," passed over mostly sprawling farmland, gobbling up an abandoned barn, a shed, and a few trees like pawns on a chessboard. The worst tornadoes are only as bad as the land they twist, and as Pat and his team looked at the radar, they knew it was unlikely to change course. Bigger towns would stand little chance, and on this night, The Beast was coming for kings and queens.

Coffee, Grits, and Motorcycle Grease

The modest clapboard-and-brick bungalows on Oak Street in Mayfield, Kentucky, were fronted by porches that became extensions of living rooms, festooned with tchotchkes that read like bumper stickers to passersby. There was the home with the calico cats that crouched around a few chipped ceramic bowls of food each morning. Across the street was the house with a Confederate flag thumbtacked into the white siding next to the front door, and a few doors down was the redbrick home with a Black Lives Matter sign taped to the front window. The cultural divide had been slowly growing along the fringes of this quiet Americana for years, shaded by the trees the street was named after.

To say Kevin Reed *chose* 419 Oak Street would imply he had some measure of control in his life. The truth is, at least according to his mother Nellie, Kevin was a coin toss away from homelessness. After graduating high school in 2000, he embarked on a journey where the only thing that was guaranteed was never hearing from him. "I worry about that boy," Nellie would tell friends—and really anyone who would listen. "Sometimes I think he has a 'problem.'" The extent of that "problem" wasn't clear. On the good days, Nellie thought he suffered from depression. On the bad days, her mind took her to the kinds of drugged-out corners she saw in television crime dramas. In the darkness that comes from silence, the mind plays evil games.

In reality, Kevin just struggled to find his way in a world that didn't give people like him a free pass. He never went to college, and his path after high school never took him too far from home. His love for his mother and his struggle to make a paycheck, let alone live paycheck to paycheck, created an invisible fence around the geography of his dreams. While his mother worried about him living homeless in another state or maybe even another country, Kevin was usually only a few miles or towns away living with a girlfriend or on his own, quietly plotting some kind of future, whatever that meant. A friend of his on Oak Street told him about 419, knowing Kevin could get a deal on rent if he helped tidy up the crumbling place. Kevin was good with his hands and loved the pursuit of fixing broken things. His latest plan was to open his own motorcycle repair shop. He liked the freedom of riding down empty country roads—the speed and that rumble—and he was good at taming mechanical horses both on the pavement and in their oily barns.

In the spring of 2020, a few months after its previous tenants moved out and when most people were on lockdown because of the Covid-19 pandemic, Kevin moved into 419 with his latest girlfriend. Neighbors were quick to notice he was the quiet tinkering type. He felt most comfortable listening to others and chose his words carefully when asked to respond. He wore a simple uniform of jeans, a T-shirt, and tan jacket, and often covered the portion of his head above his eyelids with a dark beanie. The only hair he left exposed was his beard, a thick pepper carpet with a sprinkling of salt around the chin. The house at 419, like all the other numbers on all the other walls he lived in, was just a temporary launching pad for him. And so Kevin never bought that welcome mat, never placed chairs outside on his front porch, and never turned the dusty white walls of his 788-square-foot slice of the South into the physical representation of what pulsed through him. The house at 419 Oak Street wasn't the kind of space you saged. Move-in day had a sterilizing scent of bleach.

He was outside in his yard toying with a broken motorcycle when he first noticed the premature dusk casting eerie shadows in the distance. He figured he had another hour or two before it began to rain.

The news at the time was warning about possible tornados, but Kevin didn't have cable and, even if he did, wasn't the type to tune in to the five o'clock local broadcast. He also left his phone inside. Not many people called him, and he had little interest in speaking to the few who did. As he continued his work, he made a mental note to remember to put the bike in the shed before heading in for the night.

Forty-five miles northeast of Mayfield, Becky James was in her home planning the following day's menu, also unaware of The Beast brewing nearby. In the small town of Dawson Springs, life seemed to spin around Becky and her popular restaurant, Ms. Becky's Place. Those who visited Dawson's tiny business district knew what time it was by the smell. The buttery, spicy, floral aroma of coffee, grits, and fried eggs wafted through the air beginning at five a.m. By noon it was meatloaf topped with ketchup, and when the small hand of time struck five p.m., it was the briny musk of catfish. If you followed the scent, and in the tiny community of Dawson it's really not that hard, you'd end up at a crowded parking lot with, depending on what time it was, a line waiting for a table inside.

Ms. Becky's Place was Dawson Springs's only greasy spoon and also the town's only true landmark. Sure, a few old buildings made up main street. They were the brick-and-mortar remnants of what early founders hoped would be a booming destination for tourists looking to relax in the naturally hot water the town sprung from. But most people associate Dawson Springs with Ms. Becky, and if her place was open, odds were you'd find the steel magnolia behind the register or in the kitchen or busing tables or serving her family recipes to a flood of hungry "usuals" she knew by name because she grew up with them. How long has she known Dallas Hudgens, whom she called "the Trouble Maker"? "Well, I am seventy-one years old," she liked to say. Here relationships were measured by the entirety of one's life on Earth, and payments were taken in the form of cash, credit, or IOU. A little note next to the register served as a reminder the reverend owed a few bucks for orange juice. She had told him not to worry about it. "The juice is on the house." But he insisted on paying next time he was in.

Becky was born and raised in Dawson Springs and spent the early part of her career working at a small florist across the street from the wood-sided restaurant then called The Place. She used to drop in for a morning cup of coffee or a lunchtime sandwich and fantasize about owning it. It was a dream that had sizzled since childhood, when she would watch her mother grab whatever ingredients she had in the pantry and whip together plates of crispy chicken, creamy potatoes, and fresh-caught fried fish. She loved the culinary work of theater that transformed their small kitchen and dining room into a stage. What she remembered most of all was the bewitching power each meal had to bind the entire family, no matter how busy they all were. Every night at six o'clock, a spell was cast. Here, surrounded by food, they shared the kinds of stories that defined their lives and would be told for generations to come.

Becky was nearing retirement when the "for sale" sign appeared in the window of The Place one morning. To her husband Eldon's surprise, she walked in and bought it. Eldon knew his wife's boundless energy was not chained by the hands of time that held so many others back. In only a few weeks, Ms. Becky had her name on a sign outside, and inside, people gathered around the kind of meals her mother, and grandmother before her, used to make. In the rare moments of silence that briefly struck between the morning and afternoon rushes, Ms. Becky would sit wrapping well-worn silverware in paper napkins and think about Eldon, who'd passed away shortly after her rebirth, and how proud he would have been. And when she took stock of her ingredients after the customers left and all but one light was extinguished for the night, she'd think of her mother and the magic she always managed to pull out of her pantry every afternoon before dinner.

Ms. Becky inherited her mother's ability to make anything from scratch and to bring the community together around her recipes. On the night of December 10, she made plans for a fried salmon special the following day.

Red Lodge

I pulled into Red Lodge with my windows down. The storm that brought me and my team here had passed, and the American flags that lined Broadway in preparation for next month's Fourth of July parade hung still. It was 85 degrees outside, and the occasional breeze carried the soapy scent of the lilacs that had blossomed from the rain and painted the town purple. The signs of destruction—the images that covered the front page of the *Billings Gazette* I picked up at a rest stop along the way—were nowhere to be found, and the streets were dry and empty of people. I was the grand marshal in a parade of one. It crossed my mind that maybe we were sent to the wrong location.

Covering environmental disasters is always a moving target, especially in the early aftermath when the vast and often foreign geography of an impacted region makes it hard to locate and fully assess the damage. Carbon County, Montana, where Red Lodge is located, has only about 10,000 people spread around an area the size of Rhode Island. There could easily be significant damage no one even knew about. It's also not unusual for early news reports to mix up smaller towns or give the dateline of where the report was filed and not where the actual event unfolded. Wire services like the Associated Press and Reuters, as well as forecasts, social media (when available), and local news reports all provide a rough bull's-eye when police departments and firehouses aren't answering their phones, but I've found it's always wise to leave room for error. If the GPS says it will take two hours to reach a place, I usually add an extra hour—if not longer—knowing the location we ultimately end up may not be what we initially set out for.

After driving a few blocks down postcard Broadway, I pulled over on a side street to look at the map as I chewed on a black Twizzler from a bag I picked up earlier along with the paper. Surely the photos I had seen were from another town. I was scanning X and Instagram for recent posts from residents and first responders when my producer Anam Siddiq called.

"Hey, are you here?" I asked as I chewed on another Twizzler.

"Yeah. Drive to 17th and Broadway. We pulled over in front of a roadblock." She and my other producer, Christian Duran, found the damage we had come here for.

"Okay. I'm just off Broadway now. It's a cute town, isn't it?"

"It is. And I haven't been 'hate-crimed' yet," she joked.

Anam was a practicing Muslim and wore a hijab, which she knew could make her a target, especially in a country where, increasingly, just being a journalist can attract unwanted attention. She had sprightly eyes, a quick wit, and a talent for getting people to talk, and gave not a damn about what anyone thought of her. I admired that. I, on the other hand, did everything I could to hide what made me different when entering foreign communities. "Straightening out my gay" was an old habit developed while reporting in the Middle East, but one that first took root as an intern at the network news magazine I grew up watching with my parents. A producer there, who knew I was interested in reporting, told me I had a "lovely smile" and would make "a great flight attendant." Two years later, when I was officially working as a reporter and *not* a flight attendant, an assistant news director in Milwaukee pulled me aside and told me I needed to "rein in the fagginess." I left that place as quickly as I could. Later in my career I was also denied a job at one network because the main anchor at the time didn't like the way I "tracked." It's the coded language of bigots that I've mastered over the years, and of all the bigots I've met along the way, he probably surprised me the most. This was the guy I looked up to. Working for him was the dream I had spent eight years chasing at that point. A producer later told me the same anchor only ate McDonald's whenever he traveled overseas for assignments. Don't get me wrong, nothing beats a Big Mac, but having one in Brussels instead of moules frites? Or a McRib in Munich over sauerbraten? Clearly this guy made some shit decisions in life. And yet, radioactive words can have a long half-life.

Every now and then, when I enter a small town, I still imagine people staring at me like I'm Patrick Swayze behind the wheel of his yellow 1967 Cadillac DeVille convertible in *To Wong Foo, Thanks for Everything! Julie Newmar.*

"Jonathan Vigliotti?" a man walking down the street said loudly.

I looked up from my phone, confused. "Yeah, I'm Jon."

"Oh my God. My husband and I watch you all the time. He's not going to believe this."

We talked for a few minutes, and as I drove off, I started to let my guard down a bit. As I got closer to 18th Street, I notice sandbags scattered in front of a few business doors. By the time I reached my meet-up point, the bags had formed what looked like a stone wall that stretched for blocks along either side of Broadway. I also recognized some of the buildings that I'd seen in the *Billings Gazette*, photographed surrounded by angry water. The river had since receded but left behind a smooth layer of silt on the street, a clear marker of where the surging water had trampled and where town officials worried it could again, soon. "Apparently it's supposed to rain more in the next few days," Anam said as she unpacked her camera from the trunk of her rental car. "And can you believe how hot it is?" The air, pushed by the breeze, was so warm it insta-dried the beads of sweat on my arms, leaving behind small patches of salt.

There were also boulders, the size of car tires and even bigger, lying all around. Broadway looked like a dried-up riverbed. Anam, Christian, and I set off on foot for a group of people gathered around a truck in the middle of the street a block away. They looked like volunteers or first responders. Either way, it was a good bet they'd point us in the right direction. "You came to the right spot," a man wearing a baseball cap, T-shirt, and faded blue jeans said after I introduced myself. "I'm Tom, the fire chief," he said. He was leading the effort to assess the damage, clean up the town, and prepare for another round of high temperatures and rain. "It's a dangerous combination, especially after a late-season snowstorm recently blanketed the mountains. Do you have some time?" he asked. "You should join us."

———

Tom Kuntz had seen a lot in his twenty-seven years as the Red Lodge fire chief, but what happened on June 12, 2022, was never thought possible.

He was an East Coast boy who silvered in Montana, and when you looked at him you got the sense that, aside from the color of his hair, he never really aged. He had the calming nature and athletic appearance of a guide leading a backcountry hike.

Tom took me around some of the worst-hit areas of Red Lodge and told me the story of the night his town baked in the hot air and then flooded.

"Sunday night was an absolute crazy night. It was one of those incidents where you get the first piece of information—like in this case it was initially a call of a broken gas main—and every minute that followed, it continued to get worse and worse," he told me as we walked toward the still-roaring Rock Creek. "The question was how high will [the water] rise and when will the levels go down? At that point the flooding wasn't terrible, and originally we thought the water-flows would start to recede, but I had a call with the National Weather Service and they said they were seeing continued stream flow increases and it would likely continue until three in the afternoon on Monday."

In hindsight, had that information been correct, Red Lodge would have been lucky, but a flow indicator placed upstream—the one that measured and monitored water levels—had been washed away in the current. With this critical early detection system gone, Red Lodge had no idea of the growing disaster until it arrived downstream. "The old indicator measured up to ten feet, I'm told. The river crested at over fourteen. No one ever imagined how much damage the heat and rain could cause," Tom said as we inspected the fallen section of the creek wall. He then gestured to the snow-covered peaks in the distance.

"That's where it all began. Everyone's talking about the record rain, but the main cause of the flooding was the snow. Now what's incredible is we actually had below normal snowfall and below normal snowpack for the season," Tom said. All of that changed though two weeks earlier, during a rare Memorial Day cold snap when more than four feet of snow fell. In some areas, snow drifts were 15 feet deep. A single square foot of snow, depending on its density, can contain up to three gallons of water, which in mathematical terms-for-dummies like me is a fuckton when you consider the hundreds of thousands of acres that were blanketed in and around Yellowstone. "Had we gotten those levels in the winter we would have been fine because [the snowpack] would have melted slowly over time as the temperatures slowly increased," Tom said, describing nature's seasonal frozen reservoirs, which are critical to replenishing water tables. "But instead, it fell practically in the summer. So then after that crazy late snowstorm, we had a rapid warm-up period—I'm talking unusually warm temperatures for this time of year. That snow started to soften and melt, then we got rain on top of it." The valve to this frozen reservoir broke open. "All of that brought the water that was in that snow right out of the mountains and into town."

We had now made our way from the shattered creek wall to the Rocky Fork Inn. The historic building sat on the creek at the entrance to town and was named after the Rocky Fork Coal Company, whose mining of coal in the late nineteenth century led to the development of the small community that, in 1884, grew into the Red Lodge of today. According to the Chamber of Commerce, the Sundance Kid robbed the Red Lodge Bank in 1897, and Buffalo Bill and Calamity Jane were among the colorful guests at the Pollard Hotel down the street. "The advent of strip-mining in southeastern Montana in the 1920s signaled the beginning of the end of the Red Lodge coal boom . . . and in 1943 an underground explosion killed 74 men at the Smith Mine in Bearcreek four miles east of Red Lodge, devastating the community and effectively ending coal mining," the Chamber of Commerce wrote on its website. "Not to be outwitted by the national economy, Red Lodge denizens came up with a worthwhile alternative to coal: Following the Great Depression, locally produced bootleg liquor—or 'cough syrup'—replaced coal as the town's lucrative export." Eventually, tourism would tame this wild west.

"There's a lot of history here," Tom said as he looked at what little was left of the Rocky Fork Inn. It was so badly damaged it would likely be demolished. As Tom explained, at the height of the flood, the water surrounded and even flowed through the building, which acted as a net, catching all the debris carried in the current. Boulders, tree limbs, and trunks poked out of every shattered window and doorframe like hay in a scarecrow. "I was here and watched a log—a tree, an entire tree—get shot up through the guard rail, and it stripped all the branches off the hundred-foot-tall tree. That's when we started putting down sandbags. Anything to try to stabilize the area. We had tons of citizens helping out. Hundreds of people in the middle of the night came out to try to redirect the water back into the creek, but we quickly realized things were going south fast so we then focused on door-to-door evacuations in areas we knew would be flooded." Crews with the fire department, along with more than a hundred volunteers, waded through the rising water to help rescue homeowners. "Those volunteers saved lives that night, because had they waited any longer, the water would have been too dangerous to cross through."

Not only was there the risk of drowning in the current, it was so strong it also carried massive boulders, which explained the six-foot-tall mounds of rocks that had been piled up along 19th Street. "All of these rocks were originally in the creek and carried downtown. We used bulldozers to pile them all together. You should have seen it earlier; we've probably taken a hundred dump truck loads out of here today. We'll do the same tomorrow. That boulder right there gives you a sense of the power of this water."

Tom was now inspecting a round boulder at least three feet in diameter. "You and I together couldn't lift one of the smaller boulders, let alone one of these, and tons of these larger ones were being shot up out of the river and literally carried down through the stream. They pounded the river walls and ripped apart the soil. We lost about twenty feet of land in one area, a whole area full of trees and picnic tables. The sound of those rocks echoed through the entire town for hours until the water receded," he said as we entered one of the hardest-hit neighborhoods. The rocky current was strong enough to wipe out entire streets and a car had disappeared in a massive sinkhole. Water still formed moats around homes. The sandbags that had been used in a last-ditch effort to control the flow of water were of no use.

"This is not a community that's used to filling sandbags. We don't have floods! One of my commissioners was joking at the beginning of the season—he was counting, doing a sandbag inventory, and wondering why we kept so many sandbags here because we don't have floods. Not only did we use every sandbag in our entire county, we've been bringing in thousands and thousands from surrounding communities and the state. I think it's a world of extremes. A year ago, our mountain was on fire. Our town was literally evacuated because of the threat of a wildfire. This place was bone dry. You could have walked across this creek and not gotten your feet wet a year ago. And so we were super dry, we were hot, we were windy, we had a fire start and that fire moved very quickly. A year later, today, we're wet, we have the highest stream flows that the area has ever seen. It's all connected. Losing so much vegetation last year also magnified the destruction we're dealing with now, surely. Don't get me wrong, the late-season snowfalls and rain were enough to cause catastrophic flooding, but there's nothing holding all this land together, so when rain does fall, it's like an explosion. So many roads are blocked off by landslides. It really is a world of extremes right now and the shift in seasons is unpredictable. As we see things we've never seen before, you wonder, where's the end of seeing something you've never seen before?"

HWY 89

The glossy yellow line freshly tattooed on the porous concrete was meant to divide two-way traffic, but on this day there were no cars and I used it to test my balance as I followed my guide, one foot in front of the other, to the roadblock ahead. As we walked along abandoned Highway 89, it occurred to me that the last time someone had this stretch of road all to themselves was likely when it was built 150 years earlier. That's when the gates to Yellowstone National Park, the park Highway 89 cuts through, opened to the public for the first time. In the late 1800s, 89 was just a flat path of dirt carved into the park's canyon curves about 20 feet above the Gardner River. The technique spared the development of Yellowstone's sweeping forests, filled with steaming geysers, hot springs, and wildlife like bobcats, wolves, bison, and grizzly bears. The highway helped usher in a flood of tourists, and word of this natural wonder traveled quickly. Until recently, an estimated one million cars drove down the highway every year, but on this day its fate, along with that of the park itself, was uncertain.

"This is one of the most pristine ecosystems in the world," my guide, Yellowstone National Park superintendent Cam Sholly, told me as we passed a series of orange hazard cones, "and even it's impacted by climate change." He was wearing his park ranger uniform of green pants and a light gray short-sleeve button-up shirt. He had cropped brown hair that faded to gray around the ears and a welcoming, cheeky smile. Much of the boulder-filled floodwater that pummeled Red Lodge and other gateway towns around Yellowstone's northern border originated inside the park. Rain runoff and snowmelt poured down its famed yellow cliffs and flowed into the Gardner River, which rumbled violently like a gutter in a midsummer monsoon. The park's two main northern roads, including Highway 89, were shredded and now closed. When they would reopen was unclear. "The river was so intense—was so powerful—it shaved the entire side of the canyon, taking parts of the highway with it," Cam said as we came to where the road abruptly ended. From our perch, I could make out pieces of the shattered and sunken road. Its yellow line glistened under the current of the river and looked serpentine as the water raged where 3,000 cars a day used to drive. "The total amount of lost pavement is less than a few miles, but they're in some of the hardest to rebuild and critical areas of the park," he said. "We're weighing a number of options for how to rebuild."

I met Cam to discuss those options, and none of them were ideal. To repair just two miles would cost an estimated $1 billion and take about five years to complete. "If we go with that option, we have to make sure we're doing things right. I think it's important that we don't just spend money to repair the damaged sections of road. This was a one in five-hundred-year event. I don't even know what that means. The same event could happen next year. It could happen in ten years, so what we don't want to do is make a bad investment that is compromised again in a short amount of time. It can't just be about rebuilding. We've got to reengineer. We've got to be cognizant of climate change in everything we do these days."

The other option was cheaper and quicker but more invasive. "We're also considering moving the highway along the ridgeline," Cam said, looking up above to a tree-lined plateau. Already crews were working on widening a hiking trail there, which would provide a temporary path to the park and the Yellowstone town of Mammoth Hot Springs, which had been cut off after the storm. Either option would require the largest construction effort in Yellowstone's history.

Native Americans called the terra in and around Yellowstone "Land of the Burning Ground" because of its hydrothermal vents. Tribes including the Blackfeet, Flathead, Crow, Kiowa, and Bannock lived and hunted here for hundreds of years before the first explorers arrived. In 1805, a man named John Colter had been journeying with Lewis and Clark to the Pacific when he broke off to join a party of fur trappers to map the wilderness. When Colter returned two years later with stories of bubbling cauldrons, fountains of scalding water, and cloud-piercing mountain peaks, fellow explorers didn't believe him. "Colter's tales of fire and brimstone would seldom be accepted by the people with whom he shared his tales of adventure. In fact, a region of land along the Shoshone River that today is marked by mostly extinct thermal features is known most appropriately as 'Colter's Hell'—a name that started as a joke by Colter's disbelieving audiences but is now a mark of respect for Colter's incredible journey," wrote the United States Geological Survey.

Yellowstone remained largely a mystery until 1871, when the federal government sent geologist Ferdinand Hayden along with photographer William Jackson and landscape artist Thomas Moran to explore and document the area. The photographs and paintings from that expedition were so spectacular, Congress moved to protect 2 million acres, sight unseen. In 1872, President Grant signed Congress's bill into law, creating the United States's first national park. The Yellowstone National Park Protection Act designated the region a sacred space protected "from injury or spoilation, of all timber, mineral deposits, natural curiosities, or wonders within." The National Park Service would be Yellowstone's guardian.

Sixty-two other national parks have since been created in America, with Yellowstone remaining the most visited. Collectively, the United States's national parks span 84 million acres and are among some of the most protected land in the world. But in a global greenhouse that has no borders, what's being spewed into the air in more developed corners of America and around the globe is also altering the atmosphere in these treasured spaces, with impacts that are easier to monitor in such pure and relatively controlled settings. The NPS calls this land a living laboratory for evaluating climate change, and the results from decades of data are clear. The park service itself admits that nearly every park under its purview, Yellowstone included, has been impacted in some way. "Temperatures are rising at national parks across the country . . . these changes can affect the composition, structure, and function of entire ecosystems, and alter habitats for plants and wildlife."

The buildup of greenhouse gasses has caused Gates of the Arctic National Park in Alaska to heat up more than 5 degrees since 1950. Its permafrost, a kind of permanently frozen soil that many native towns are built on, has thawed into quicksand. California's Joshua Tree National Park has warmed by more than 3 degrees in the same time, resulting in a 39 percent drop in rainfall, and 1.3 million trees, about 25 percent of the parks population, were destroyed by a wildfire in 2023. Rangers in Montana's Glacier National Park say it's only a matter of years before all the glaciers are gone. Everglades National Park in Florida is being flooded by rising sea levels that are threatening endangered species of plants and wildlife. In Yellowstone, the average annual snowfall has decreased by more than two feet since 1950, and the snow that does drop on the mountain range is arriving later in the season. More frequent spring heat waves deliver both hot air and rain that rapidly melt these frozen alpine reservoirs. This snow and ice is then flushed down valleys, into communities like Red Lodge, and through streams and rivers before emptying into the Gulf of Mexico or the Pacific Ocean, instead of thawing out slowly over time and hydrating the land like a dripping irrigation system.

The NPS is now drafting ways to adapt their hardest-hit parks to climate change, and they've had help from the federal government. For all the criticism President Trump's environmental policies have received—and most of it is well deserved—in 2020, he flooded the National Park Service with money. The Great American Outdoors Act allocated $900 million a year to the Land and Water Conservation Fund and provided up to $9.5 billion over five years to help maintain the country's national parks. President Joe Biden's $1 trillion infrastructure bill, passed in late 2021, set aside an additional $1.7 billion to specifically help upgrade roads and bridges and support environmental adaptations. It all sounds like a lot of money, until you do the math.

According to National Park Service records, the agency manages more than 12,600 miles of roads nationwide, 40 percent of which were in need of repairs according to a 2019 park study. In Yellowstone, the price tag for roadwork in 2019 was $1 billion. That was *before* the flood hit. Highway 89, the one that would cost an estimated $1 billion to repair, wasn't even on the park's to-do list. In 2020, after the first round of funding was approved, John Garder, the senior director of budget and appropriations at the National Parks Conservation Association, a nonpartisan advocacy group for the NPS, said the money was "helping stem the tide, but certainly not enough."

Similar financial hurdles are also hitting America's concrete forests. In New York City, erosion from floods and inadequate drainage led to a 38 percent spike in sidewalk and street sinkholes in 2021 compared to the previous year, according to the mayor's office. City leaders linked the problem directly to climate change and said repairing the sunken holes to last another fifty to one hundred years would be expensive but save taxpayers money in the end. They pointed to a report from the National Institute of Building Sciences that found that, for every dollar a community invests on climate adaptation, they save six dollars on needing to rebuild again. But as our extreme elements outpace humanity's ability to adapt to them, not all problems can be solved by throwing money at them. While asking for more funding for sinkhole repairs, Rohit Aggarwala, NYC's chief climate officer and the commissioner of the Department of Environmental Protection, warned money would only help so much because they were quickly exhausting all available engineering solutions. "The issue right now is we don't know exactly what we would do with more money that would systematically reduce the likelihood of sinkholes," Rohit said at a city council meeting on the topic in 2022. Protecting modern-day

Rome and places like Yellowstone (Mother Nature's Notre Dame) from ecosystem collapse will require more than just dollars. Fortunately, there is another lifeline, but it too has started to fray.

The Clean Air Act (CAA) was passed in 1970 and is considered by legal experts to be the most powerful environmental law in the world. Overseen by the EPA, which was established around the same time, the CAA monitored and restricted harmful air pollution in American cities. The act initially targeted gasses including carbon monoxide, lead, and nitrogen dioxide, but expanded over time to include carbon dioxide as concerns over global warming grew. The act gave the EPA the power to place hefty fines on companies that emitted these gasses at toxic levels and is partly credited with keeping annual emissions in the United States relatively stable since the 1990s, despite the nation's population growing by nearly 80 million people. Even so, there's still plenty of room for progress. The U.S. remains one of the highest emitters of CO_2 per capita in the world, and collectively, the global community emits around 36 billion tons annually, according to Global Carbon Project. That's a more than 200 percent spike since the 1960s. While the CAA was seen as a blueprint for developing nations, the landmark policy hit its own major roadblock in 2022 when the Supreme Court limited the EPA's ability to regulate carbon emissions from power plants, which alone are responsible for around 25 percent of the United States's total CO_2 output each year. The vote was 6 to 3, with the courts three liberal judges in dissent saying the majority had stripped the EPA of "the power to respond to the most pressing environmental challenge of our time." They weren't exaggerating. Our air controls everything from summer highs to winters lows and has the power to throw Earth's natural cycle into a tailspin of extreme and unpredictable reactions.

The same radicalized air that caused a freak spring snowstorm followed by a freak steamy downpour in Yellowstone was also linked to a series of heat waves that killed 339 people in Arizona. The summer of 2022 was the state's deadliest on record, and the summer of 2023 was the hottest, with temperatures in Phoenix exceeding 110 degrees for an entire month. Overnight lows never dropped below 95. ER doctors showed me how they treated a wave of patients suffering from heatstroke by slipping them into body bags full of ice and water. My team and I also rode along with paramedics who were injecting patients with ice-cold IV to rapidly cool them off. It's not just heat. This destabilized air is also responsible for transforming a typical blizzard in Buffalo, New York, into an apocalyptic winter storm that entombed entire neighborhoods in walls of ice and snow. Dozens of people died, and many of the victims were found trapped in their cars. The deep freeze quickly engulfed the city like a fast-moving wildfire, and you could almost hear the crackle of ice forming when looking at the pictures of the frozen still life.

"These events have been called freak acts of nature, but they're not. So much of what we've witnessed has been building up over time. It's probably easier to notice in a national park," superintendent Cam Sholly said as we headed back to more stable ground. "There's no question climate change is occurring. People can talk and debate the cyclicality of it. The fact is it's happening, and we've got a long way to figure out what steps are necessary to ensure that we're adapting properly. There's not just one answer to this problem. It's going to require a holistic approach."

"Get to Shelter NOW"

By 8:30 p.m. on December 10, 2021, The Beast had already passed over largely unpopulated land in Fulton and Hickman Counties in Kentucky but was on track to hit Mayfield, about forty miles away in Graves County. Pat and his team couldn't believe their eyes. They had never seen a tornado so big and with so much power. They had hoped it would veer away from populated areas, but its path became terrifyingly clearer with each passing minute. At 8:55 p.m. the Paducah field office made the call. The twister would hit Mayfield head-on. There was no way around it. At 9:03 p.m., Pat typed a tornado emergency alert and beamed it to every form of communication possible. "ALERT*** Heads up Mayfield Kentucky. Tornadic Storm moving your way, could arrive by 9:30 p.m. Be ready to shelter immediately!" He was so nervous his fingers were sweaty and he nearly forgot how to spell Mayfield's name. Another alert was sent at 9:10 p.m. "If you live in or near MAYFIELD, you need to be underground if at all possible. Get to shelter NOW!" Residents remembered the haunting crescendo of sound as the Emergency Broadcast System electrified landlines, cell phones, televisions, and radios. They were the final metallic notes of life before.

"Are you seeing these alerts?" the man asked when Kevin finally picked up his phone. It was his neighbor down the street. He sounded worried. By now the lightning in Mayfield was flashing every second. The mile-wide tornado was so close, those in the neighborhood couldn't even make out where it began and ended. Kevin remembered looking out his window at the marble texture of what he thought was the sky, not realizing it was the outer wall of The Beast bearing down. He was alone in the house and decided to run across the street to ride out the storm with his friend. "There was safety in numbers," he thought. Within a minute of his friend's door closing behind him, The Beast hit.

Kevin could see the walls breathing—bowing in and out—and the floorboards lifting in violent contractions. A cold sensation came over him as the storm sucked the warm oxygen out of his lungs. It felt like he had stuck his head into a freezer . . . or more aptly, out the window of a plane. It was hard to even speak. His hollow words were ripped from his mouth and spit 30,000 feet up in the air. The pressure caused his ears to pop. Time lost all value in this moment, and something strange came over him, like the kind of serenity that sinks in when you know no matter what you do and what happens, the end result was never in your control. Kevin grabbed hold of the door; the pressure from his tight grip made his knuckles gray, but the rest of his body eased as it quickly adapted to the rhythm of chaos. For a second his feet were lifted by the wind, which felt like fingers wrapped around his ankles. The blood in his body now rushed to his head and carried in its crimson tide the vibrations of his pounding heart, which rattled his heavy eardrums. *Boom-boom, Boom-boom, Boom-boom.*

Kevin was in Mother Nature's womb, and he was about to be pushed out into an unfamiliar world. At some point—it felt like a few minutes—the force of the wind began to subside. Kevin's hands tingled as the blood started to reenter the small vessels in his fingers, now no longer so tightly gripped around the doorframe. The worst of The Beast had come and gone. Kevin was in the final, slippery seconds of deliverance.

The Beast only took three minutes to pass over Mayfield, but Kevin collapsed in exhaustion like a runner breaking the tape on his first marathon. He was too weak to move and too out of breath to speak. "You okay?" his friend finally asked, also out of breath. His raspy words helped Kevin regain his focus. He finally let out a heavy sigh. "I'm good, man," he said, looking around like it was the first time his eyes had seen the world. Incredibly, he and his friend and his friend's home were still standing. Windows were busted, part of the roof was ripped off, and rain was leaking through massive cracks, but the screeching stopped and Kevin could once again hear the sounds of the world he was ripped from—like the sound of water leaking in from the roof. On any other day, that plunking would be a nightmare—the sound of money for a repair his friend couldn't afford—but on this day every plop, drip, and drop was proof of a past life. About thirty minutes later, still shaken, Kevin went outside to inspect the damage. The air was so thick with mist and the dust of pulverized land that the beam from his flashlight could only cut through about ten feet. In that moment, Kevin learned to fear Mother Nature.

As Kevin pieced together what was left of his ecosystem, ten feet at a time, Becky was about to set off on her own roller-coaster ride as The Beast approached. She was sitting at the dining room table when she got a call from her neighbor. "They're saying the storm is coming right for us and we should take cover in a basement," the voice on the other end said. Becky's house was the only one in the neighborhood with one, so she called her children who lived across the street and told them to hurry over. "Don't waste time now. We don't have time on our side," she warned before hanging up the phone. Eight of her family and friends ran over. They all quickly disappeared downstairs and packed into the windowless room her late husband Eldon had used as an office. His collection of John Deere memorabilia lined the walls.

As The Beast moved directly over her house, violently shaking every inch, Becky was most scared by how quiet everyone in the room got. That silence was louder than the storm. Then there were the eyes of her children and grandchildren, glazed over by terror as stress pumped through their bodies. Everyone was so still as to look almost lifeless. They weren't just taking shelter, they were *hiding* from the storm, as if it was alive and searching for people. They could hear it ripping off the roof and breaking through windows. It was trying to pull open the door, which they now held on to, using all their weight. Three minutes later, The Beast gave up and moved on. Becky and her family sat quietly for a few minutes, paralyzed by fear and worried, as irrational as it was, that The Beast would return. They emerged from the office—first slowly cracking the door before opening it wide—to find several walls and half the roof of Becky's home were gone. Thick pools of sludge were dripping in like drool. The monster had come close to grabbing them. Becky closed her eyes and thanked God. She wasn't worried about the house or the rest of the stuff. When you come that close to death, you learn right away those things don't matter.

Kevin and Becky don't know each other. More than 47 miles of bluegrass, red maple, American beech, and bald cypress divided them. Their social circles—that of a forty-year-old soft-spoken, hardworking motorcycle mechanic and a seventy-four-year-old firecracker who runs her own small-town diner—had never Venn diagrammed. But on the balmy winter night of December 10, 2021, the burnished edges of their Southern lives, along with those of tens of thousands of strangers, overlapped and were welded together by just three minutes of chaos—the time it took for The Beast to pass over. That night, Mother Nature stirred a cauldron of warm and cool air that turned 180 seconds into a lifetime for Kevin and Becky. They'd been through a lot during their collective billions of Mississippis on Earth, but what happened that evening not only erased their towns, it made the impossible possible, and ignited the kind of fear that changes ones understanding of the world.

By midnight, December 11, the National Weather Service had issued a total of 149 tornado warnings across portions of nine states, including Kansas, Ohio, and Indiana. The supercell storm covered more than 550 miles and spit out 66 tornadoes in all, with The Beast registering as not just the largest of the pack, but also the largest in Kentucky history. Pat and his team tracked it for more than 160 miles. *One hundred sixty miles!* While the worst of the storm was over, he worried about the aftermath. Hundreds of thousands of Kevins and Beckys had been in its path, and he knew alerts only saved lives if they were seen and taken seriously.

He ultimately took comfort in knowing his team had given people twenty-seven minutes' notice to take cover. In the past, people would receive an alert only a few minutes before a twister hit, if they were lucky. Radar and the science behind what triggers twisters have improved drastically over the years.

Twister Chasers

Generations earlier, eighteen-year-old Roger Jensen and his father looked more like ghostbusters than storm chasers as they barreled down dusty country roads in the family's white Chrysler DeSoto. Small rocks and other debris dinged the car's undercarriage like loose change in a drying machine. Chasing tornadoes in Fargo, North Dakota, is an unusual father-son outing, made all the more remarkable considering it was the summer of 1953 and nobody had done what they were doing before.

Roger first fell in love with weather as a child. He was especially drawn to the light show that unfolded in the first moments the translucent edges of a storm cloud covered the sun, projecting the turbulent atmosphere on the open grassy plains below. As a teenager, Roger spent months squirreling away money from odd jobs on the family farm so he could buy his first camera from the shop on Main Street and capture Nature's fleeting tantrums.

This one grew like all passions, as snapping clouds soon spiraled into tailing twisters. His Kodak Pony 828 exposed every kind of tornado before science had even given them names. There were the thin, braided "rope tornadoes"; wider, longer-lasting "cone tornadoes"; even wider, more destructive "wedge tornadoes"; and the supercell thunderstorms like the one over Searcy, Arkansas, that could produce all three, known as "multivortex tornados." Soon science wouldn't just name these monsters, but also measure their strength and assign corresponding rankings, ranging from an EF0 to EF5, the strongest, capable of producing winds more than 200 miles an hour and pulverizing brick-and-mortar buildings right down to their foundations. The photos from Roger's adrenaline-fueled, mid-century pursuits ushered in a new field of research that later led to scientists such as Lou Wicker and Steve Smith taking even greater risks to try to get pictures from *inside* the storm.

In the 1980s, during the months of April and May, you'd likely find Lou and Steve eating gas station burritos next to a rusty roadside pump, waiting for Oklahoma's volatile storm season to spit out its latest twister. The duo were part of NOAA's elite Severe Storms Laboratory, and their fieldwork, much like man's ability to study tornadoes at the time, was based largely on a science of "ifs." In those days, live Doppler radar didn't exist. It was their job to analyze weather data from the morning and forecast where a tornado-producing thunderstorm could develop. *IF* the storm did form and *IF* that storm did spit out a tornado and *IF* they were close enough to it, then they would not only chase it, they would get out in front of it, hauling new technology in the bed of their truck that would need to be placed in the twister's path. (. . . The '80s were fucking wild.)

The TOtable Tornado Observatory, appropriately nicknamed TOTO after the terrier in *The Wizard of Oz*, was a fifty-five-gallon, barrel-shaped instrument that held hundreds of small devices capable of taking meteorological measurements, including pressure and humidity. When placed in the path of an approaching tornado, a hatch door on top of TOTO would open, allowing the storm to suck the device and its sensors into the vortex, giving researchers a 3D image of how the twister looked and a better understanding of the kinds of conditions ideal for turning air into atmospheric muscle. At this stage, how TOTO *should* perform was only theoretical. As it turned out, using hours-old information to predict where a tornado would form precisely enough to safely place the device in the oncoming path of the deadly storm was just as difficult and dangerous as it sounded. Getting swept up in a tornado aside, it didn't help that TOTO, a tall metal structure, was a perfect lightning rod.

Lou and Steve had made dozens of unsuccessful drops that they wrote off as practice for April 29. On this day, the tornado they hoped to capture was a perfectly formed EF2 cone moving on a relatively stable track of flat farmland. Wind speeds were approaching 150 miles an hour, enough to tear roofs off homes and uproot trees, but fortunately this slice of land was clear of most obstacles. The conditions were, by all accounts, perfect. A cloud of dust trailed Lou and Steve's Dodge as they tumbled after the weather system on high. Once they hit the mile threshold of the storm, they ground to a halt, shifted the Dodge into park, and left the engine running as they carried out a sequence of synchronized steps. First they lowered the tailgate, then positioned the ramp, removed the 350-pound TOTO, unlocked its top hatch, repositioned the ramp, slammed the tailgate shut, and peeled out down the unsettled trail of dust that traced their earlier route. All of this was done in about forty-five seconds. They celebrated from a distance as they watched, through binoculars, their tornado carve a path right over TOTO. It was a direct hit . . . followed by a gut punch when they went back to inspect the site. TOTO had failed to launch. Instead, they found it tossed on its side.

NOAA retired TOTO in 1987, but it did have one successful launch a few years later. The Severe Storms Laboratory was the inspiration behind the movie *Twister*. Director Steven Spielberg renamed the instrument Dorothy, and in the final scene her hatch door opens, releasing hundreds of sensors into the storm—the kind of historic victory Lou and Steve imagined. But even at a distance, their field observations remain part of the earliest scientific record on the impact wind shear, humid air, and terrain have on tornado development and behavior. It's a record that still has a lot of catching up to do.

"We've been studying tornadoes for decades, but they're such quick and risky events that don't provide a lot of time and opportunity to gather data," meteorologist Harold Brooks told me. "Tornadoes are inconsistent and short-lived, and most of the data from them has little value." Harold spent time in the late '80s chasing tornadoes with Lou Wicker and later helped envision that fictional launch of TOTO as a consultant for the film *Twister*. Today he's the senior scientist at the Severe Storms Laboratory, where advances in radar technology have helped meteorologists like Pat Spoden understand more about tornados in the last ten years than the previous forty combined. But while technology and field research are enabling meteorologists to predict the perfect conditions for creating a tornado, Harold says it's still difficult to forecast exactly where one will touch down and where it will track.

What is becoming clearer is the role climate change is playing in fueling a deadly new pattern in America's tornado season. Harold coauthored research, published in 2018 in the science journal *Nature*, that looked into the impact our changing climate was having on where tornadoes formed. Tornado Alley—which cuts through Texas, Louisiana, Oklahoma, Kansas, South Dakota, Iowa, and Nebraska—used to be a twister's traditional path of destruction. But as these states' climates became drier over decades, Harold observed a measurable shift east into the increasingly wetter, warmer, and more populated mid-South. "These are areas that have more people because they were considered relatively stable. You never used to see this kind of weather there." Harold has given this emerging front line on the map—which includes towns like Mayfield and Dawson Springs—his own nickname: Tornado Fatality Alley. A 2023 study published in the *Bulletin of the American Meteorological Society* goes one step further, linking our warming climate to an increase in those deadly supercell systems capable of producing multiple tornadoes at once (like the storm that formed over Searcy). "Results reveal that supercells will be more frequent and intense in future climates, with robust spatiotemporal shifts in their populations. Supercell risk is expected to escalate outside of the traditional severe storm season, with supercells and their perils likely to increase in late winter and early spring months under both emissions scenarios . . . These results suggest the potential for more significant tornadoes, hail, and extreme rainfall that, when combined with an increasingly vulnerable society, may produce disastrous consequences," researchers found.

Places once considered safe havens, and even those protected as national treasures, are now threatened by our shifting air, and the laws created to hold back our radicalized elements—one of the main tools the U.S. has implemented to pump the brakes on the changes to climate that enable catastrophes like The Beast—have been weakened.

The massive path The Beast carved through Kentucky could be seen from space and found on the ground by following makeshift signs that directed victims to free food and water. One had "FEMA" written in large red letters. There was a tent pitched on a street corner with a poster that asked passing traffic "Need to talk?" The signs grew in volume and the number of people Walker and I saw as we entered town shrunk the closer we got to the epicenter. Soon we were passing once-towering brick buildings that now littered the street like fallen Jenga games. The few traffic lights that still worked flashed red for drivers that didn't exist. Mayfield, a small town of about 9,800 people, was empty. By the time we reached the historic downtown, there was nothing left untouched. Century-old buildings that had stood the test of time had crumbled under weather that was weaponized by it. We pulled over to take a photo of the old welcome sign that was painted years ago on the side of a half-standing beige brick building.

MAYFIELD

MORE THAN A MEMORY

The Land Between the Rivers

In 1817, Mayfield, Kentucky, was part of an area known as "the Land Between the Rivers." Woodlands here teemed with wildlife. Tributaries veined through lush grasslands. Fertile soil gave birth to sunflowers, peas, beans, squash, pumpkins, melons, and corn. When white settlers first learned of this Eden, they immediately wanted it, but the Indigenous Chickasaw people had already raised it as their own.

These early settlers, with dollar signs in their eyes, must have thought the Chickasaw's modest way of living was a waste of potential. According to the Chickasaw Nation's archives, this homeland was dotted by pitched structures that disappeared into the thick vegetation, a camouflage that acted as both means of survival and a sign of respect for land that, in return, gave them everything they wanted: fresh water, vegetables, fruit, meat, hides, and fur. In the summer months, families gathered in a simple Toompalli' chokka', rectangular tent-like structures that allowed the breeze to easily pass through thin walls made of woven grass mats. In the winter, the Chickasaws lived in a sturdier Hashtola' chokka', which was partially sunken into the ground and covered with clay for better insulation. These homes surrounded a council house, where every morning leaders convened to discuss important tribal matters and events—and there was no matter more important than what was discussed on October 19, 1818, the date Andrew Jackson, then a famous judge, accomplished general, and political leader, arrived in the Land Between the Rivers.

Jackson, along with former Kentucky governor Isaac Shelby, was appointed by President James Monroe to negotiate a treaty with the Chickasaws. Monroe wanted this slice of the map, and returning home without it was not an option for Jackson and Shelby. Representing the Chickasaws were brothers Levi and George Colbert and Chief Tishomingo, who served under Jackson in the War of 1812 and was awarded the silver medal by President George Washington. When the men emerged from the council house, 8,500 square miles of Chickasaw territory had been signed over to the United States for the equivalent of $11 million in today's money. The so-called Jackson Purchase was seen as a win-win by both parties at the time, but the Chickasaws, who resettled deeper into the South, would soon have seller's remorse.

By the following year, settlers began flooding the Land Between the Rivers, tearing down the Toompalli' and Hashtola' chokka' and clearing the forests they were nestled in to build cabins, farms, and churches. The natural riches found in the Land Between the Rivers were exactly as advertised, and the settlers' hunger for more land continued to grow. In 1830, then-President Jackson approved the Indian Removal Act, which used militia might instead of smooth-talking checkbook diplomacy to grab more ancestral Indigenous land in the Southeast in exchange for less desirable federally owned territory west of the Mississippi. Around 60,000 Native Americans—men, women, and children—were forced from their homes and had to travel by foot, wagon, and horseback more than 5,000 miles along the "Trail of Tears" to their newly designated Indian reserves in areas that make up today's Tornado Alley. Many suffered from exposure, disease, and starvation, and thousands, including George Colbert and Chief Tishomingo, died before reaching their destinations. Historians call the Indian Removal Act a genocide and academics say Adolf Hitler later used it as a model for his own extermination of the Jews. In Nuremberg, while addressing what he called "the Jewish problem," the Führer declared the Volga River in Europe "must be our Mississippi," documents University of

Georgia professor Claudio Saunt in his book *Unworthy Republic: The Dispossession of Native Americans and the Road to Indian Territory*.

The town of Mayfield rose from America's holocaust along the Mayfield Creek, which was named after an early settler who drowned in it. By the mid-1800s Mayfield was known in the region for clothing manufacturing. The Mayfield Woolen Mills opened in 1860 and the successful Curlee Clothing men's suit store opened in the early 1900s. Ambitions of growing into a clothing capital were scaled down in the 1960s when competition from factories up north ate into regional demand. Boomtown Mayfield transitioned into a sleepy suburban oasis.

At a recent high school reunion, former classmates reminisced about spending Friday nights hanging out in the parking lot of the Mayfield Shopping Plaza, the one with the Piggly Wiggly and Little Caesars. These days it's hard to find any sign of Mayfield's ancestral past, though every so often someone will pluck an arrowhead or two from the crisp bubbling creek the town was named after. Today, Mayfield is perhaps best known for its grand Victorian courthouse and prominent clock tower, whose own transformation runs parallel to that of the land. In 1823, a few years after the Chickasaw left, it was built using logs at a cost of $139. In 1834, it was replaced with a two-story brick structure costing $5,400. The building was destroyed in the Civil War and rebuilt in 1888, where it has stood ever since. The hands of the dial ticked clockwise, give or take a repair or two, until they stopped for good on December 10, 2021.

———————

Like a kitchen doorframe with the dates and heights of children penciled on it, Mayfield's clock tower measured time and the growth that unfolded during the passing of it. A vibrant town rose around the courthouse in the seconds, minutes, hours, days, months, and years since it was built. The chiming of its bell at noon could be heard in small businesses along the main street, like Red's Donut Shop, In Its Time

Antiques, Sue Ann's Dance Factory, Duncan Clinic Pharmacy, Fowler's K&N Root Beer Drive-In, and also inside the updated and immaculately maintained historic homes that lined nearby South Seventh. Residents on Oak Street lived too far away to hear the chiming, and in a sense were forgotten by time. Homes here weren't the type for renovations and didn't have sprawling front yards with soccer nets and campaign signs staked into the grass during election years. The bungalows on Oak Street were shells for people to outgrow—like hermit crabs.

Shortly after 9:30 p.m. on Friday, December 10, The Beast severed the brick octagonal clock tower from the body of the courthouse, leaving a cavernous hole in the roof that exposed what was left of the tower's spiral stairwell. If you were to climb those fractured steps you'd have a full view of the mom-and-pop wasteland below. More than a century of growth had been erased down to the wooden floorboards. And what few buildings did stand were illusions. The Methodist Church's facade of neo-Gothic columns survived but wasn't holding up much. Part of the roof was blown off and wooden pews were tossed around with Bibles still in the seatbacks. Broken glass crumbled under my feet as I snapped a few photos. The red velvet seats in an auditorium at the American Legion down the street remained bolted in place, but the brick wall behind the stage they faced had been ripped off in a perfect rectangle, revealing destruction on the other side. It looked like an empty screening of an apocalyptic movie.

Oak Street had escaped the worst of the storm, but that is not to say there wasn't severe damage. In a neighborhood with preexisting cracks, it was just harder to see the new ones that had formed. The black and blue tarps placed over missing sections of roofs almost blended in. Homes were still standing after the storm, but some barely. "It's hard to believe, but there are people who are still living here," Brian Montgomery said as we walked down the street, a lanyard with several laminated badges dangled from his neck. He had short gray hair and a walrus mustache (that's the official barbershop definition) that framed the corners of his mouth. He spoke with a gentle drawl that reminded me of Billy Bob Thornton. "There's a lack of shelters in this area and some people don't have the means to even find one or get to one." Brian was the field leader for FEMA's Disaster Survivor Assistance Program, and Walker and I joined him as he and his team canvassed Oak Street looking for survivors to help register for assistance. The bold yellow "FEMA" lettering on the backs of their dark blue jackets made them look more like enforcers than help. And maybe that's why, when the team of men and woman holding clipboards showed up on porches, some residents were reluctant to peek through their windows, let alone open their doors.

FEMA was officially created by President Jimmy Carter in 1979 in response to a series of natural disasters in which local and state authorities requested needed backup. FEMA's modern capabilities—medical care, search and rescue assistance, support with cleanup and recovery, temporary housing, and federal grants to rebuild—have become well known as disasters strike larger and closer to come, but FEMA got its bearings in fits and starts. Back then, local and state agencies were mostly able to get by on their own, as nature's pace and damage picked up slowly over the next two decades. In 2002, following the September 11 attacks, President George W. Bush consolidated FEMA and twenty-one other organizations, along with their budgets, under the newly created U.S. Department of Homeland Security. Had the frequency and volatility of storms remained at pre-2001 levels, this reorganization may have worked, but the White House failed at that time to recognize and protect against Nature's own radicalized elements. To park FEMA under an agency whose main priority was counterterrorism was a shortsighted decision at best. Hurricane Katrina, and the inundation of New Orleans within just a few years' time, proved that point quickly.

The Katrina response's mistakes were many, but can be best summed up as bureaucratic bullshit. A murky chain of command required FEMA field teams to first get approval from layers of supervisory staff back in Washington before taking action. Victims in desperate need of help were stuck in the highest-stakes equivalent of a DMV line, and countless lives were needlessly trapped in a deadly web of red tape. Estimates put the number of dead at 1,392 people, many passed while waiting for help according to a report published by the American Medical Association. In 2006, Congress passed the Post-Katrina Emergency Management Reform Act that established FEMA as a distinct agency with its own direct line to the president. In theory, this should have helped speed up the approval and delivery of aid. Still, the agency struggled to keep up with the radicalized elements, as later exposed by 2017's historic hurricane and wildfire seasons. Resources were spread too thin, and even having the White House on speed dial wasn't enough to get help quickly. The criticism continued. The Disaster Recovery Reform Act of 2018 gave FEMA expanded autonomy and a bigger annual budget—nearly $22 billion. In what is becoming a common theme, the changing genetics of our biosphere and its ecosystems continued to outpace the agency's progress.

In 2020, FEMA responded to 314 disaster declarations compared to just 30 in 1980, according to their online records. In forty years, annual wildfire response jumped from 3 to 82. Hurricane disasters went from 1 to 30, and severe storms including tornadoes increased from 7 to 22. The cost of the damage is also sobering. In 2020 there were 20 "billion-dollar" storms? Well, forty years earlier in 1980, there had been just 3.

That trend continued. The multivortex storm that spit out The Beast and dozens of other tornados on December 10, 2021, became the nation's most destructive of its kind, causing an estimated $5 billion in damage, according to FEMA. Most of the ninety people killed that day lived in Mayfield.

"We've seen it all, including having to help people arrange funerals. Most of the time we're meeting people who are overwhelmed by the destruction and just don't know what to do. The damage from storms like this has the power to remind us all to appreciate more what we've taken for granted in life," Brian told me as we approached 419 Oak Street.

That was the first time I stepped foot on that sterile front porch. The home had been tarped over, not unlike several others on the block—a sign someone cared enough about the place to lend it cover after the storm. I couldn't see any movement through a warped slat in the blinds, and while it seemed pointless to knock, the team did anyway. "FEMA," a woman said with a warm, inviting tone. They stood still for about a minute, the amount of time they knew it typically takes a person to fully assess their unannounced visitors from the other side. The door cracked open and a set of brown eyes framed by glasses peeked through.

"Yes?" Kevin calmly said to our group of five.

"We're with FEMA and are here to provide assistance," a man holding a clipboard responded. "How are you doing?"

"I'm doing good. Real good," Kevin replied in a soft baritone as he now stood on the porch. "I lost a bunch of things, but I'm still here. That's the main thing." I don't know if he believed they could help, or just appreciated the conversation . . . any conversation. The house had lost part of its roof and there was water damage, but for now he was getting by. He was spending his days fixing bikes he'd found that had been damaged in the storm, but wasn't all that sure how much longer he'd stick around. The eerie silence of the place had become unsettling, he said, especially at night.

"Why did you stay?" I asked after introducing myself.

"I don't have anywhere else to go, I guess. I haven't lived here all that long but now I'm connected to this place. It's my responsibility. I walk around and check in on the neighbors' places. There's some worry about looting. I'm slowly trying to pick everything up," he said calmly. "It's not going to pick up itself."

And it was true. The active cleanup we had seen closer to the center of town had not reached these battered outskirts yet. FEMA assistance arrives faster today than ever before, but there's still plenty of red tape to cut through. Even so, every agent in the field is a hero. They're often the first to arrive and the last to leave a disaster, and sacrifice months away from their own families to help strangers. Fighting Mother Nature can be a lonely and sometimes thankless battle. "But to see these communities come back, and for some it can take decades, but its inspiring to help give people a second chance. And at least for me, it's a reminder that we can overcome this," Brian told me.

Kevin gave the FEMA team his social security number and confirmed his name and address before disappearing back into the house. "I've been feeding the neighborhood cats," he said, returning with a bag of food. "There's no one else around and they were looking hungry." He then went to the side of his house to set up his tools for another day of tinkering. The bike he'd forgotten to bring in that night was cracked up and dented from being tossed into the side of the house, but overall the damage was nothing he couldn't fix.

"Have you thought about how you recover from something like this?" I asked.

"I'm not so worried about what it will look like for me as much as I am for the next generation. By the time it gets back to how it 'was' I'm probably not going to be around. But the only way we're going to recover the right way is if we stop thinking in terms of 'how am I gonna benefit right now?' Nah, this is my world now," he said, looking around at the destruction. "Hopefully we can rebuild it so the next generation never needs to pick up any pieces again. Mother Nature's ruthless. You got to respect every aspect of her and I do. I really do."

All Hands on Deck

FEMA agents arrived in Red Lodge on the same day we did and quickly set up their base camp at the high school on 16th Street to begin registering residents for federal assistance. They were a welcome sight in town, of course, but in the aftermath of most disasters, the quickest form of relief often comes from volunteers—mostly local, but in Red Lodge, for one, a swarm of people seeking to provide aid came from out of state. A cafe owner from Cody, Wyoming, set up a tent on Broadway and gave away free meals for breakfast, lunch, and dinner. Then there were teams of men and women who went door to door to help clean up and dry out damaged homes before the mold set in. They were a lifeline for impacted families, and no task was too small.

I met Bri Beekman, a shy eight-year-old girl with long brown hair covering a freckled face, a few days after arriving in Red Lodge. Volunteers were washing mud out of the hair of her Barbie dolls. She had an impressive collection that was now spread around the backyard. The dirty ones were soaking in a series of clear plastic bins, and the freshly cleaned dolls were sunbathing along the curved ledge of a trampoline. Their dripping synthetic hair dangled over the side. About four or five other people were moving in and out of the house placing damp items in the sun and moldy drywall and carpet in the front yard where a heap of trash had been collecting. There were also a few other children, neighbors I think, who were chasing each other outside, wound up by the excitement that came from having so many unexpected visitors in the neighborhood. One of the children, a boy no older than six, told me his dog liked to play soccer. He kicked a ball and I watched as the shepherd blocked it with his nose and walked away with it in his mouth.

Bri was in her bedroom when the floodwaters began seeping in late Sunday night, and called for her mom, who was upstairs and had no idea what was going on. Her father, Dan, a truck driver, was on the road for work and was also unaware when his wife called to tell him the basement and first floor were taking on water. "I felt helpless as a father and a husband," he told me as we walked through his muddy backyard. The floodwater was still pooling in some areas. In all, more than 300 homes and businesses were damaged or destroyed by the flood, but Dan considered himself and his family fortunate. Down the street, a home built along the creek had the earth pulled from underneath it. The two-story cabin was now leaning sideways in the rapids. The water, while receding, was still raging through its submerged windows and a door. "It's all about perspective," Dan said. "We have it bad, but at least we still have our home."

Repairing his home was the next challenge. "I mean, it would be nice if we had flood insurance, but it didn't make sense. Floods never happen here, and financially it was too expensive for such little risk. Luckily, the community is fantastic. We literally could not do the demo, do the reconstruction, without the help of everybody. So already tons of labor, tons of resources have gone into helping us clean up. You know, we're down to the studs already. It's actually going to happen a lot faster than we thought. It was so daunting at first." His voice then got lower so Bri couldn't hear. "Honestly, I was hesitant to talk to you because I'm concerned the state is now going to require us to get flood insurance, which we can't afford," Dan said, referencing the federal government's new Risk Rating 2.0 plan, which requires any designated flood-prone home with a federally backed mortgage to secure flood insurance. "The whole system is flawed. It already happened to us in Colorado where we just moved from. We had a once-in-a-hundred-year flood and overnight our home, which was considered low risk before, was raised to the highest risk and we were required to get flood insurance. It was so expensive we were forced to leave. We actually factored in flooding when we bought this home. I mean, it never flooded here in the history of this town. We're running out of high ground." What politicians in New Jersey and New York had worried about following Hurricane Sandy, had arrived in the Rockies. As Dan described, under the federal government's current flood plan, even homeowners who go out of their way to minimize risk were being penalized by a system that not only fails to identify threats to the ecosystem but is struggling to adapt communities to it now. Meanwhile, the costs of flooding only continue to mount. A single inch of water in a home can cause roughly $25,000 of damage, according to FEMA.

I left Dan so he could get back to cleaning, and ran into the fire chief near my car. Tom was making the neighborhood rounds. "Only eighteen people have flood insurance in all of Carbon County," Tom continued. "There was never a reason to be concerned. This isn't even considered a flood zone. A lot of people are kind of on their own. FEMA can provide up to $30,000 per household, but those grants are hard to get and many people are walking away with far less. Then there's the issue of finding someone who's available to do the repairs. We're grateful to the agencies here, don't get me wrong, but there's only so much they can do. We're not the only people in the country they're helping these days. The storm has showed all of us the power of community. There are easily hundreds of volunteers from the town and even other states doing stuff in a few days that would take months to fix," he said. You could see their work everywhere you looked. Hoses that led to basement windows were spitting floodwater out on the streets. Wet clothes were hung on trees to dry out. Electronics and damaged furniture were placed in front yards for disposal. Hammering and drilling and the buzzing of saws and the sputtering of generators could be heard for blocks. "It's a long road, but we're going to come back from this ready for the next 'once in a thousand-year flood,' even if it happens again next year," he laughed as we said goodbye.

It was late in the afternoon and my flight home was the following morning, which meant I was reentering the in-between—that state of existence before returning to the luxuries of a life that, just a few days earlier, the people I was reporting on took for granted. In these final hours of every assignment, I am cautious not to open my big mouth, aware that stories of returning home can be triggering for those who can't.

Ms. Becky's Place

Wednesday's special at Ms. Becky's Place, down Broadway in Dawson Springs, included a choice of *one* meat: pork chop, liver and onion, roast beef, country-fried steak, or smothered chicken, with a choice of *three* sides: pinto beans, mashed potatoes, beets, green beans, slaw, cabbage, green peas, fried okra, or apples. All made from scratch. All for $6.79.

Walker and I arrived in the middle of the lunch rush. To walk through the doors was to be transported back in time to Dawson before the storm hit. Every red-laminated table was topped with stuffed plates of piping hot food and surrounded by people—more people than we'd seen in the previous days filming the destruction along The Beast's twisted path. Seventy-five percent of Dawson was damaged or destroyed, and 15 people were killed in the town of 2,452, according to the mayor's office. I recently read a study that said the average American has nine "close friends." In tightly woven towns like Dawson, that number is likely higher. Everyone knew someone who didn't survive.

The chitter-chatter and metal-against-ceramic clatter inside the diner drowned out the ringing sound of loss that haunted survivors. "You must be Jonathan," Ms. Becky (in the flesh) said in a softened Southern accent. Her eyes were mischievous, and her hands and arms were loaded with plates of fresh food. "I'd like to think I know a camera crew when I see one. Make yourself at home," she said with a laugh as she continued on her kitchen-to-table path without skipping a beat.

She was a Southern belle still chiming from table to table despite the cracks.

"Here ya go . . . now be careful, the plates are hot," she said as she delivered her food, before heading off to the next table. "You want that burger with pickle, onion, and mustard?" she asked as she wrote down the order on her pad. "Now don't go missing out on your sides, you get three." Every time I looked at her, she was on to a new task, always with a smile on her face. "Ms. Becky's Place, good morning," she was now back the counter answering the phone. "How would you like that cooked? Great. Well, we appreciate you calling. See you in a few minutes."

She delivered the food, took the orders, convinced customers not to skip out on their sides, cleared the tables, washed the dishes, filled up and dropped off glasses of water and soda and juice, rolled silverware, answered the phones, seated customers, combined tables for larger groups and split tables apart for smaller ones, arranged the fresh homemade pies in the glass display, flipped burger patties in the kitchen, and helped customers settle their bills, which was her favorite part because it gave her time to chat.

"How's the family doing? Have you started on demolition?" she asked one customer. "No, we're still in the picking-out phase. We've got a lot of work ahead," he said as he grabbed his receipt.

"Ms. Becky, I heard you lost the house," another man said while pulling a $10 bill out of his pocket to settle his tab. "Yep, the home and everything. I'm living in a donated trailer now. But nobody got hurt. That's all that matters," she said, and you could tell she meant it. After all, a woman who rolls the silverware and helps wash the dishes when she's got a staff that could do it truly loves every moment of life.

This lunchtime dance both on the dining floor and back in the kitchen carried on for hours without pause until about four p.m., when Becky, her waitresses, and cooks—all women except for one male dishwasher—gossiped and poked fun at each other and their journalist guests in a rare moment of calm before the dinner rush.

"If you film me with one of those cameras of yours I may have to kill you. I do not like cameras," Sam the cook said while asking if anyone had a clean shirt for her to change into. "For just in case I change my mind," she said with a wink.

"You remember my name?" a waitress teased as she walked by me. Denise, a half-pint raspy-voiced flirt with a black pixie cut, had been asking me the same question all day for no clear reason other than she must have intuitively known I'm terrible at remembering names. "Now don't you go forgetting my name . . . I'm gonna quiz you all day," she threatened with a smile on her face.

"Denise, how do you even know who you're talking to when you can't see over the counter?" another waitress named Debbie said playfully. She was sipping on a fountain soda, reading the local paper—the same paper where I read about Ms. Becky reopening her restaurant after being closed for several weeks because of storm damage. "Ms. Becky lost her house, the dishwasher lost his house, and I lost mine," Debbie told me. "Right after the storm I remember thinking, at least I have this place to come back to, but then I got a call from Sam and she told me this place was gone, too, and then it was really bad for a little bit. To lose everything you know, everything that is familiar, it's scary. I was very depressed. I didn't even want to move. You ask yourself, how are you gonna get back on your feet when the ground was ripped up from under you? I remember how quiet everything was. There was just silence. But not the kind of silence that comes from being tired or people not wanting to talk. It was silent because there was nobody around. It was really bad. Then we got word that we were still standing here. Ms. Becky's had survived. I knew I had a home to come back to and was relieved.

"Sometimes it takes losing something to appreciate what you had even more. I always appreciated this place, but I didn't realize how much everyone else did until we reopened. There were people lining up outside the door waiting to come inside. The other night, we put the pies in the display—Ms. Becky makes the best pies—and this guy pulled into the lot and came through the door and bought them all. Every last one. He didn't even ask what kinds they were. He said he'd been driving by the shop for days looking in the window as he drove by—we light the pie display up at night so you can see right in from the front window. He said it reminded him of home, and I knew what he meant. I opened up last Monday for the first time since the tornado and it was the first time I had any normalcy." Debbie slowly drew out the word "nor-male-see" in a way that snapped me out of the illusion unfolding inside the walls of Ms. Becky's and how abnormal things were on the other side. Town officials said it could take a decade to fully recover, if not longer.

The diner's reopening was Dawson's great comeback story and a reason to celebrate when reasons, those days, were few and far between. "We lost power and had some damage to our roof, but we were in good shape overall. It felt like we were closed for more like six months than a few weeks," Becky said. "It was an isolating time for the whole town. I was down in the dumps, having a pity party, but when I came back through these doors there was laughter, food, friends, and I was back up on my feet. The hardest part was the customers we lost. The ones that aren't rebuilding or who passed. I mean, this place does become a family. You get used to the people who come in every day. There's the coffee drinkers that come before the kitchen's even open, the afternoon bunch who come in for a quick bite and catch-up, and those who stay until the last light is turned off. You learn their routines. Everyone has a routine and we become part of that routine. Like, we'll have some folks who go trout fishing, and they'll be gone for a week and we'll all miss them. We know they're out there in a better place—I love being on the water—but when there's someone here every day and they're gone, you just miss them," Becky said. I got the sense she was talking about more than just fishing for trout.

A handful of lone customers still hung around at the tables, picking at hours-old plates and looking out the window at the cleanup effort underway. They had nowhere else to be, or maybe there was nowhere else they'd rather be. It was a fine line after The Beast ripped through town. The rising steam from a freshly poured cup of coffee captured the milky, late-afternoon rays of the fading winter sun. Even before the storm, Ms. Becky's was busy, but lately the traffic felt more like something you'd find on a late-spring weekend in downtown Louisville, by the university.

But closer to my home, it reminded me of O'Hara's on 120 Cedar Street in lower Manhattan. The Irish pub and restaurant is two blocks away from the World Trade Center and sustained significant damage when the Twin Towers fell on September 11, 2001. I was a freshman at Fordham University, and in the months and years that followed, spent a lot of time at Ground Zero reporting on the cleanup for WFUV, the NPR affiliate on campus. When O'Hara's reopened seven months later, it became a destination for crews working on "the pit," the name given for the gaping hole where the towers once stood. O'Hara's was a place to mourn the lost and celebrate the surviving . . . "full of good spirits and good people," as the owner put it. The dust that once lingered in the tonic and tobacco-scented air has long since settled, and today, O'Hara's is a living memorial covered in embroidered patches from firefighters, police officers, and paramedics. The bar has two scrapbooks documenting the tragedy that are available to look at upon request.

No two disasters are the same, but I have learned over the years while covering many of them that everyone has an O'Hara's—a gathering place to pick up life's pieces and acclimate to a new normal. Ms. Becky's Place was that place for Dawson Springs. Police officers, sheriffs deputies, FEMA crews, men and women with the State Historic Society, engineers with the Army Corps, families who had lost their homes and people who lost loved ones gathered here for different reasons, but all driven by the desire to adapt and rebuild. The tornado had done unbelievable damage, and looking around town, it was easy to think the planet was changing in ways and speeds that humanity can't or won't keep up with. But to spend time inside Ms. Becky's was to realize the human spirit is stronger than Mother Nature's wrath.

"You should never make plans on an empty stomach," Ms. Becky said to me as we talked at a freshly cleaned table. The Windex streaks glistened in the light before evaporating. Walker and I had spent the entire day with her, and she was convincing me to stick around for some meatloaf, which didn't require much convincing at all.

"Food is a form of love. It's a way of taking care of people. If you feed somebody, you help them. You make them feel better. It's just love," she recited as we dug into our plates. That belief had been instilled in her by her mother, grown over years, and was alive more than ever following the storm when little else was guaranteed. Now she held court, stepped fully into her position as an emotional center to a town determined to hold on. "I guess there's a lot of fear these days, and that comes from people saying Dawson won't be the same and it won't be back, but my heart tells me it will," she said. "It's like my recipes. I follow what my mother used to do but add my own touch. We all need to be open to adjusting. You have to cook with what's in the pantry."

Our conversation was briefly interrupted by an older, frail-looking man. He stood over us slightly hunched, holding a hat over his chest with his gently shaking hands. "I've been thinking for several days on exactly what you were going to have for a special tonight. They were wonderful. When my wife died, I started losing weight, and I've been thinking about those vegetables, and you had them all tonight. Thank you," he said as he slowly walked out the front door. "It melts my heart. It feels like I'm fulfilling my purpose," Ms. Becky said, her eyes glassy. Her food *was* that good, but it's Ms. Becky herself who's Dawson Springs' real special.

Sounds of Life

I spent close to a week in Red Lodge. On the first day, I survived off gas station boiled eggs, jerky, and black Twizzlers. On the last night I had a rib eye, medium-rare, with a glass of pinot noir. People were starting to crowd the streets again. I took a stroll down Broadway after settling my bill. It was 10:28 p.m. and the air was 73 degrees, according to a flashing sign in front of the local bank. A few drops of rain fell, but not the amount Tom feared would come. The river was still raging, but the sound of crashing boulders had mellowed to a rain stick—yeah, that was it—which you could still hear throughout town. I traced the path of the flood, curious to understand how it twisted, turned, and rumbled in the middle of that night. The silt left over from when the water receded carpeted the street like crushed velvet. There were still a few boulders scattered around.

I turned off Broadway and onto 15th Street, where the power was still out. It had been six days. A gas lantern backlit a man as he rolled tobacco on the hood of his mud-caked Wrangler. He was drinking what looked like a can of beer. I couldn't make out his face and he couldn't make out mine, but his shadow connected with my feet. We were two ghosts passing in the night. I then turned into the alley between Broadway and Platt, attracted to light coming from a few blocks away. A cat scattered as I walked by, and the purple lilacs swayed in the gentle breeze. I felt cleansed even though it had been two days since I showered. The light got stronger, and soon I could make out the rattling from a cymbal and the strings of a guitar. It was coming from the Snow Creek Saloon.

A group of twenty-something-year-olds was smoking outside. Their conversations blended together. I stepped inside and ordered a drink: "Scotch on the rocks, please." I stood near a pair of girls who were throwing darts and showing each other videos they took of the flood on their phones. "It was the scariest thing I've ever seen," one said. "I didn't even want to come back to town, but I needed to pick up clothes," said the other. Above them, a band played from a stage built into the rafters. The room smelled of pot and sweat, and for the first time since arriving to Red Lodge, I couldn't hear the raging river over the sounds of life. It had been less than a week since it ripped through town, and already people were reclaiming what belonged to them, even if still unsure about their future.

PART FOUR

EARTH

Father of Soil

To many of the farmhands who wore the typical denim uniform of Levi's overalls, Hugh Bennett was a most unusual sight in the field. No matter how hot, humid, muddy, or dusty it was, he always wore his woolen suit. Sometimes even a three piece one. Dr. Hugh Bennett was no man of dirt. In America's heartland, he was known as the "father of soil."

Hugh was born on April 15, 1881, in Anson County, North Carolina. His parents were cotton growers and he spent his early years on the family plantation, where the seasonal crop was so abundant it seemed money grew from the ground. The period was a time of explosive expansion for the farming industry. From 1860 to 1906, the number of American farms tripled, according to the United States Census Bureau, and millions of acres of wild land were stripped and cultivated to meet the demand of a growing population. But little was known about soil conservation and its impact on crops. Sometime in Hugh's teens, the once-fertile family farm stopped bearing the fluffy fibrous currency. Land that had sprouted what looked like acres of fresh snowfall in September went bald with little explanation. What happened to his parents' plantation led Hugh to the University of North Carolina, where he studied chemistry and geology in pursuit of a root cause and possible solution to his family's demise. Soon, he learned his parents were far from alone. Farmers across the country were reporting a sudden and inexplicable drop in crop yields.

In response to this emerging crisis, the federal government began investigating, and called on Hugh to help lead the effort. After graduating college in 1903, he joined the United States Department of Agriculture's (USDA's) newly created Bureau of Soils, which at the time treated soil as an indestructible resource. Hugh investigated declining crops all over the country and abroad, including in Costa Rica, Panama, and Cuba, and identified common themes among the disparate lands. One thing was clear: The evidence showed soil *was* destructible and easily exhausted.

Most notable in Hugh's research was the impact that removing native vegetation had on the land. When moss, grass, bushes, trees, and other plants—along with their deep roots—were ripped out of the ground and replaced with thin rows of cultivated crops, this newly tamed agricultural land exposed the topsoil to the sun, causing it to dry out, which made it harder for it to absorb water. And because these seasonal crops had shorter roots, there was less holding the soil together.

Then there was the threat from planting the same crop in the same soil season after season. Hugh discovered that what are known today as "monocrops" depleted the soil of natural microorganisms that need biodiverse soil packed with a variety of plants and wildlife to live. The tiny creatures are critical, he learned, because they break down organic matter like leaves, roots, and dead bugs, and in the process, release a fresh batch of nutrients into the soil that help plants grow. With fewer native plants and their more developed roots, the dry soil and its phosphorus, potash, magnesia, sulfur, and nitrogen were being washed away by rain and blown away by the wind. Hugh became a champion of adopting early Native American growing practices that called for rotating a symbiotic mix of crops and leaving the land in a more natural state.

Hugh would spend more than a decade warning the country of a crisis on the horizon, and the realization of that dark future brought little solace when it arrived. In 1921, the *New York Times* took note of his campaign. "[Bennett] preached his gospel at crossroads villages and church socials. Not many listened," it reported as he was in the midst of his calls for a return to a more natural state. Hugh rang the alarm again in 1928. "Soil erosion is the biggest problem confronting farmers of the Nation over a tremendous part of its agricultural lands," he wrote in a report that gained too little traction, too slowly.

Loosening soil—not unlike today's extreme heat emergency—went noticed but not acted upon, long before the crisis point arrived. But once given "life" by wind, millions and millions of tons of that soil sparked terror and entered the imaginations of the public. The black dust first rose from the ground in Oklahoma and other states like the inky smoke of a house fire, and farmers who had never witnessed a dust storm before thought something was ablaze. The clouds reached so high into the sky they turned day into night, and by the time people realized what it was they were looking at, the sharp particles had reached them.

Humans had unwittingly triggered a new kind of American weather event known as "black blizzards"—stinging storms that became so routine, the Midwest and Southern great plains, the breadbasket of America, would eventually be identified by another name: the Dust Bowl. By early 1934, an area of land the size of the state of California had been rendered useless, and mass cases of lung disease had claimed thousands of human lives along with countless livestock. "Americans have been the greatest destroyers of land of any race or people, barbaric or civilized," Hugh would later write. In 1933, as the environmental disaster became impossible to ignore, Hugh was appointed head of the federal government's newly created Soil Erosion Service.

In what was called "Operation Dust Bowl," Hugh's team enlisted farmers and ranchers across the country to transform their land. Grants of federal assistance were given to help with implementation of a series of conservation techniques. Rows of trees and grasses were planted to act as wind breaks. Land managers were trained in soil retention and crop rotation techniques designed to keep topsoil protected and beyond the reach of windstorms. Hugh also persuaded President Roosevelt to use funding from the New Deal to pay farmers directly in exchange for agreements to leave their land fallow and allow it to regenerate its nutrient load without the burden of producing crops. Over a few seasons, grain surpluses were erased and prices rebounded to levels that made agriculture a sustainable career again. In extreme cases, and at Hugh's guidance, the federal government even purchased land from farmers and ranchers outright. By 1937, nearly five million acres of barren farmland were reclaimed and permanently returned to their wild state. Today, most of it is protected as National Grasslands—there are twenty spread out over thirteen states—which share similar status to America's national parks. Swift action mattered.

By 1939, as droughts broke and rainfall returned to the great plains, Hugh had helped create a new kind of farmland that absorbed the water, shielded soil from the wind, and gave birth to a new era in conservation. His framework for agricultural management was copied around the world. In Brazil, Hugh's April 15 birthday is even celebrated as a national holiday, National Soil Conservation Day. But back in the United States, Hugh's legacy, while deeply guarded by some, has mostly vanished like dust in the wind for many.

Why else would we once again be treating soil like dirt?

Dexter Dirt

I wasn't necessarily looking for Barb Kalbach when I began to type my way through Google, on the hunt for a story I'd been wrestling with in my mind. I was thinking of an idea more than a person. I don't recall the specific words I plugged into the search engine, but the result felt predestined, like the bubbly answer in a Magic 8 Ball. Poring through a results page of headlines from small-town news articles about family farms vanishing from the map, Barb's eyes stopped my scroll wheel in its tracks. She was photographed standing in what appeared to be a plowed cornfield. Her face was framed by silver hair, Aqua-Netted into a stiff bob. I'd be lying if I said Barb didn't remind me of Mrs. Clause from the Sears department store my parents took me and my brothers to as kids during Christmas. But Barb was wearing a baby blue ski jacket and leaning against a wooden fence that marked her wintery property line. Dexter, Iowa, as pictured in winter, was a raw landscape of iced-over snow, veined by leftover wheat from the previous season. This was no North Pole, but it was another world far from mine, and I knew in that moment, as I stared at Barb through my digital one-way window, that I had to meet her face-to-face.

It was early February when I touched down and saw those veins of golden wheat glistening in snow, like the yolk of an egg bleeding into flour. The brick buildings along Dexter's main street were all baked to last, but most now sat stale, caked by fumes and dust from passing farm vehicles. I pulled over next to one in search of life and peered in through the windows below a faded "Weesner Drug Store" sign that hung over the entrance. The window was draped with a yellowing bedsheet, but through a small gap I saw what looked like the pharmacy counter and empty shelves. A calendar was hanging on the wall, flipped back in time.

"What ya taking photos of, son?" a man called out from the entrance of American Legion Post 419, the only building on the street that appeared to have visitors inside—if the cars in front were any indication. His breath, spiked well with gin from the bar inside, lingered in the 10-degree air, blurring his face. "I'm just trying to capture some of these buildings. They all look like they have a lot of history," I said, without saying much. I was in chameleon mode. "You should visit the Dexter Museum," he suggested, pointing to a small single-story brick building across the street. The sign on the door marked it closed. "That's the only place where you can see what this use to look like. A lot of these towns are disappearing. There ain't no one around here interested in farming no more." I wouldn't meet Barb until my next stop, later that day, but I knew she would have her own thoughts to provide on the state of America's heartland.

Agricultural Chessboard

Dexter, like most towns in this part of southern Iowa, is a flat landscape of cash crops and livestock boxed off by dusty roads that were first carved into the land when European settlers began cultivating it in the mid-1800s. That's around the time Barb's ancestors arrived, along with hundreds of other farming families looking to plant capital. This place was bustling then, with families scratching out their places in the sun right up against one another. Four generations later, Barb and her husband, Jim, operated one of the last family-owned farms in the whole town. They were running out of time as *factory farms* quickly picked off their neighbors one at a time on this agricultural chessboard.

"Factory farms" or "industrial farms"—controversial umbrella terms for large-scale farming operations with more than 1,000 acres and/or 1,000 head of livestock—began to dominate the board decades ago. As the corporate game plan currently goes, major companies either outright purchase and merge smaller farms (operations that are 500 acres or less) under one warehouse-type "barn," or contract with sellout family farms, pumping money into these mom-and-pop fronts, which in return grow produce and raise livestock for them in an enforced process that resembles an auto assembly line. If one of these farms doesn't meet their quota, their contract ends. The family farms that do remain—the ones that don't sell out—are often too small to produce the kind of bulk needed to keep up with the big guys. In so many words, their days are numbered. As Barb would soon school me, factory farms not only push out family farms, they unravel entire communities, overwork the land, harm food security, and are a mounting threat that generational farming communities have been unable to keep at bay, despite their best efforts. While some small farms are teaming up and combining their land in an effort to fight off these corporate Goliaths, in towns like Dexter there simply aren't enough neighbors left to wage this rural war.

Barb and I were sitting around a small table in the kitchen as I watched her sketch Dallas County and take stock of her neighbors on a pad labeled "shopping list." CNBC was muted on a television tucked away on the counter near a machine that scrolled through the current rates per bushel for crops like wheat, corn, and soy. A "Stop Factory Farms" sign in red was pinned on the wall next to the thermostat. Barb was the bookkeeper of the family; her husband, Jim, the calloused hands.

"You have the Dennings here," she explained as she scribbled their name on one corner. "Then there's the Watts's farm and the Shoesmiths. We're here." Barb then scratched an "X" under each of their names. "All gone. That's the emptying out of rural Iowa. Square mile by square mile." She spoke matter of fact. The sound of the kitchen clock ticked in the background. And it wasn't just farming families that were running out of time. "You're losing entire towns. When the Dennings and the

Wattses and the Shoesmiths leave, that's less people that are visiting the repair shop, the equipment dealer, and buying feed downtown. That's less animals that need veterinary care. And so when you lose families you also lose businesses, and all of that means less tax dollars to keep public programs in place. We have to drive an hour away to see the doctor and the same distance to fill our prescriptions. We still have our elementary school, but if we lose that, we'd have nothing." And the odds, she said, weren't good. According to the National Center for Education Statistics, around 4,400 schools in America's rural districts closed between 2011 and 2015 because there wasn't enough funding or attendance to keep them open. By contrast, suburban districts added around 4,000 new schools in the same time period.

Barb's sketch of the modern family farm defied the glossy images of rolling green fields, bright red barns, and happy cows that plaster products on stocked supermarket shelves. Our drive to meet Jim at his workshop a few squares away helped color in her lines. "So that's the Shoesmiths' place I was telling you about. It was a great big family farm. It used to be very active. There were cattle and pigs," Barb said, pointing at a lifeless plot. The decay worsened as we drove. The Watts family's farmhouse was sinking into the ground. The roof, Barb told me, had long fallen in. And all that was left of the Williamses' farm was a hen house full of weeds. The Dennings' home on Walnut, a two-story shingled colonial with a wraparound porch, had been bulldozed. "It was a beautiful home, but when the Dennings retired, the children didn't want to take on the business." Unlike David, Dexter was no match for Goliath. And it wasn't just Barb's square, or the squares around Dexter for that matter. Two-thirds of the counties in Iowa have seen their populations decline since 2010, according to the U.S. Census. It's a trend in rural communities across the country. Since 2007, around 200,000 small farms have shuttered, according to the USDA, with the average farm size around 400 acres. Another study by the USDA found the majority of surviving farms lost money year after year. "Of the roughly 2 million U.S. farm households, slightly more than half report negative income from their farming operations each year. The proportion incurring farm losses is higher for households operating smaller farms." When land is forfeited or sold, it's often quickly added to corporate coffers, largely out of sight of the cities where most of us live and take for granted the true cost of our daily supper.

This new model of factory farming first took root by weaponizing a series of game changers small farmers were ill-equipped to adapt to fast enough. For starters, globalization has turned countries, in some cases thousands of miles apart, into neighbors fighting for the same customers. Unfortunately for family farms, these international neighbors offer much better prices. Don't like the cost of California-grown tomatoes and avocados? Mexico sells them cheaper. Don't want to pay for pricey oranges grown in Florida? Go to Brazil; more than 50 percent of orange juice in American refrigerators is shipped by specialized fruit juice tankers from São Paulo, according to a report by the nonprofit Supply Change. It's a lot like searching for the cheapest fuel on a block packed with gas stations that advertise each gallon down to the penny. Those pennies quickly add up when you're buying product by the millions. Small farms simply can't compete. And when President Trump slapped China with a tariff on steel and aluminum in 2018, the nation retaliated with their own tariffs on American agriculture. According to the Peterson Institute for International Economics, the United States's soybean exports to China fell from nearly $12 billion in 2017 to only $3 billion in 2018 as the Chinese government sourced the same products from other nations for either the same price or cheaper, and with relative ease. Make no mistake, even if a trade war ends, and a new president decides to undo a policy change, the old way of business doesn't just snap back post-truce.

Then there's the impact of unpredictable weather—droughts, floods, soil erosion, and erratic irrigation patterns linked to human-caused climate and habitat change—which takes its pound of flesh well before those market forces ever come to bear. In 2019 alone, unseasonable rain and snow prevented farmers from planting 19 million acres and destroyed millions of other acres already planted, according to USDA records. To that point, it was the worst year in more than two decades' worth of bad years. Between 1995 and 2020, farmers received $143 billion in federal crop insurance payments, most of it because of crop damage.

These days, the only way for small farms to survive this web of threats is to plant more, eerily similar to the strategies employed in the 1920s prior to the Dust Bowl. Planting more means either over-working the soil or owning more land, which—if you've been following along—most small farms simply can't afford to do, even if they wanted to. Today, factory farms are the ones doing most of the planting, and they will say they know how to manage the land safely this time. But research shows that many large-scale operations are cutting corners for short-term gains. "The soils of the American Corn Belt were once celebrated for their fertility. But industrial farming treats that fertil-ity as a resource to be tapped, not maintained," writes the Union of Concerned Scientists. This consortium of leading American scientists and researchers found industrial farming "leads to several kinds of [environmental] costs," including soil depletion, poor irrigation, ero-sion, and the loss of biodiversity. Sound familiar? These "costs" are all linked to an insatiable appetite for more land, and today, factory farms are gobbling up more land than ever before. So how much are we actually talking about? We'll get to that shortly, but first to this land grab mastered over time.

Early on in their history, factory farms acquired acreage from aging farmers who didn't have family to take over the business. But over time, corporate agriculture began to speed up the process, too hungry to wait for farming families to decide when to get out of the business on their own. Corporations have learned well how to rig the game.

The first step is usually obtaining low-interest, federally backed loans to establish a foothold in an agricultural region. Once enough land is secured, the corporate farm then floods the market with every-thing from meat and milk to produce and grain. The resulting surpluses reduce the value of commodities so much it's nearly impossible for their less-funded and smaller family-owned competition to break even.

And here's another added layer of fucked-up: Corporate farms know the government will eventually buy their surplus in an effort to stabilize prices for the smaller guys. But by the time such a stabilization does happen—and the reality is these factory farms are constantly overproducing and thus always driving down prices—the small farm can no longer hold on.

And for those who can manage to squeak by, another layer of unfair competition awaits. Entire sectors of the industry have grown around big corporate players, including megawholesalers and slaughterhouses, which put smaller main street operations known for competitive pricing out of business and treat the small farmer that walks through their new doors like a beggar off the street. It's no surprise Chapter 12 bankruptcies were up nearly 13 percent in the Midwest from July 2018 to 2019 and 50 percent in the Northwest, according to the American Farm Bureau Federation. The vast majority of these losers are the small guys. Now back to that earlier question of how much land their Goliath competitors control. We'll never truly know. Under the Obama administration, the EPA began to take stock of the nation's factory farms, but halted the effort after industry groups sued. "[The industry] has avoided any effective regulation and accountability for a long time," former EPA lawyer Michele Merkel told PBS shortly after quitting over the agency's reluctance to take action against megafarms.

This unregulated monopolizing of American agriculture explains why, even as more than four million small farms have disappeared since 1948, total farm output has more than doubled, according to the USDA, carried out largely by corporate-backed factory farms who hollow out main streets and overwork the land. "You can only abuse the land so much before it stops giving back," warned Barb. Factory farms, though, are showing no sign of slowing down. While these megaproducers only account for around 10 percent of all American farms, they produce around 80 percent of the food in supermarkets, according to the USDA. Put another way: While around 90 percent of America's farms are classified as "small," they only provide about 20 percent of the country's food.

"This doesn't just impact us. Less competition also means less variety, which ultimately hurts you," Barb said, pointing at me as we continued our tour by car, coming to a stop sign that looked as if it hadn't seen traffic in years. "When more of our food comes from a single source, what happens to our food security if there's a recall and the factory farm is forced to shut down? We're getting to the point where a few corporations control who eats and who doesn't."

And the threat is even greater than that. As the family farmer disappears, along with them goes the generational knowledge of the earth that corporate farms don't have. Family farmers know the weather and its impact on everything from seeds to the soil they're planted in. Studies show family farms also use less pesticide and fertilizer, and produce higher yields per acre than factory farms. The research suggests if more land was owned by family farms, it would be healthier and produce more. The lessons from the Dust Bowl, while a century old, are part of every small farmer's DNA. "Factory farms, especially the ones that raise livestock, are often automated facilities operated remotely with minimal staff, and don't have the same knowledge and certainly not the same community investment. A lot of the equipment is run by a computer," Barb said as we passed land that had recently been purchased by a corporate farm. "So, in some cases, you've got a guy in an office somewhere in Timbuktu, not even on the farm, running the land remotely. Now you've lost the innate understanding of the land. What happens when all that wisdom is gone for good? Most farmers aren't in this line of work to make money, and that's a good thing, because most don't. We're just trying to look after the land for the next generation." They sounded like the fading words of farmers from more than a century ago.

Creampuff Pioneers

According to research by the Kennedy Center, as many as 400,000 of the Dust Bowl's 3.5 million eco-refugees flooded California alone as they escaped harsh weather and unlivable conditions, straining the state's urban centers and its already limited natural resources. Ironically, local farmers at the time were at war with the desert city of Los Angeles, which had recently built a system of aqueducts to divert water from agricultural land in the state's central Owens Valley. More people meant less drops to go around. California wasn't the only unwelcoming state. Nationwide, refugees were met with resistance, and in some places even assurances from the president of the United States himself weren't enough to survive.

As part of Franklin Delano Roosevelt's New Deal, the federal government purchased thousands of destitute farms in the bowl and moved families to unused land across the country. Nearly a hundred of these so-called resettlement communities were established from Florida to California, but only one location was classified as "experimental." "Alien" was maybe a more fitting word, because nowhere else in America can you find farmland that grows under a midnight sun and in the shadow of a glacier.

The federal government lured families to the remote Matanuska Valley Colony with promises of fertile land and seasonal bounty. The minerals in the soil, they were told, could grow incredible crops. But in the early 1930s, when Dust Bowl colonists set out on trains and ships for their own "New World," they might as well have been on a *Mayflower* headed for the moon.

The Matanuska Valley, about 45 miles north of Anchorage, Alaska, is a wrinkled and rumpled land of Triassic collision, formed when Earth's tectonic plates crashed into each other hundreds of millions of years ago. The resulting slow-motion wreck pushed crustal pieces up from the Pacific Ocean, creating what would eventually become the Alaska Range. During the last ice age, a massive glacier two miles thick covered most of the region and carved out veins of valleys as it began to naturally retreat some 20,000 years ago. This clashing of land also parallels the cultural collision that unfolded across the entire state in the creases of what was ancestral Indigenous territory.

Alaska's land grab first began back in the 1700s when a small group of Russian settlers arrived and claimed the region as their own, even though 100,000 natives, including Ahtna, Dena'ina, Inuit, Athabaskan, Yup'ik, Unangax̂, and Tlingit, were already living there. Those settlers struggled to survive in the harsh conditions, and in 1867, Russia sold the territory to the wealthier and better equipped United States, which promptly stripped all native groups of their rights to the soil and began to exploit it for oil and gold.

By the early 1900s, Matanuska Valley was a mostly non-native world of beauty and extremes. The Chugach Mountain Range was snowcapped year-round, and its rugged base was blanketed by dense evergreens patrolled by black bears and wolves and 7-foot-tall territorial moose. The snaking Matanuska River was the valley's glistening silver necklace. It's only fitting that this geological marvel became home to one of the last great pioneer movements in American history . . . and it only makes sense things didn't go exactly according to plan for everyone involved.

The city of Palmer was first established in the valley in 1916 as a blink-and-you'll-miss-it whistle stop on the Alaska Railroad's branch line, which had been recently laid down to transport workers to the newly developed Chickaloon coal mines. It's safe to say not many people stepped onto Palmer's wooden train platform to explore. In fact, by the early 1930s, ahead of the colony's arrival, only a handful of non-natives had built homesteads in the region.

Alaska at the time was still a U.S. territory whose true value remained unknown only because tapping into its rich resources meant risking life and limb. As the federal government saw it, planting farmers there would help ease the agricultural crisis in the lower 48, while opening the foreign land to exploration and settlement. In less than a year, the U.S. government surveyed the land and then spent $5 million (the equivalent of more than $100 million in today's money) to relocate and subsidize around 200 farming families on about 260,000 acres of land. Candidates, according to news reports from the time, were chosen mainly from Minnesota, Wisconsin, and Michigan, because the government believed the winter weather in those states closely resembled Alaska's harsh arctic conditions and would therefore be survivable. They were mostly wrong.

The hastily cobbled together exploratory mission in America's last frontier was criticized before it even launched as a failed "social experiment" that both exploited the desperation of dust-blown families and eradicated the few native tribes that had managed to hang on to their way of life. The disapproval was prescient, because when colonists started to arrive in early May 1935 after traveling weeks by train and ship, many reported back horrific conditions. The farmland they were guaranteed was covered by thick forest. Basic infrastructure, including lodging, was nonexistent, and the equipment needed to clear the land and prep the soil—equipment that was promised to the colonists—was nowhere to be found. Families had no choice but to sleep in train cars until transient workers could complete a temporary tent city.

The lack of adequate food, bathrooms, and medical care contributed to an alphabet of diseases that swept the colony, including polio, measles, chicken pox, and pneumonia. Once the trees and thick vegetation were cleared from the land, farmers discovered the soil also wasn't as advertised. Much of the earth inside the Matanuska Valley was covered with coarse gravel and massive rocks left behind by that ancient sheet of ice, and a lot of the soil was frozen year-round. That permafrost made planting, irrigating, and draining impossible. And the land that *was* successfully cultivated faced a number of hurdles, including a small growing window of just a few months. There was no room for error in an industry where there was already little room, even under ideal growing conditions.

On June 16, 1935, a group of desperate colonists sent a telegram directly to President Roosevelt that read: "SIX WEEKS PASSED NOTHING DONE NO HOUSES WELLS ROADS INADEQUATE MACHINERY TOOLS GOVERNMENT FOOD UNDELIVERED COMMISSARY PRICES EXORBITANT EDUCATIONAL FACILITIES FOR SEASON DOUBTFUL . . ." In only two months, twenty-five families left Alaska before even breaking ground. In just five years, half the colony dissolved. The national press skewered the failed farmers, calling them "creampuff pioneers." But little coverage was given at the time to the twenty or so families that hung on, and the spoils they eventually found in the earth. As it turned out, Alaska's nearly twenty-four-hour summer Sun acted like a steroid when harnessed correctly. While the state's arctic weather gave way to a short growing season of only 100 days compared to around 270 in California, the endless light provided scientifically proven photosynthetic muscle. Everything from Brussels sprouts to broccoli just kept growing and growing, but nothing compared to the mammoth vegetable that swelled by dozens of pounds a day and weighed as much as a washing machine, with a leaf span of more than 5 feet.

The Einstein of Produce

Alaska's official state flower is the alpine forget-me-not. The state animal is the moose. And if the last frontier had an official vegetable, it would be the giant cabbage, which would make Palmer, population 6,094, the state's capital. Everything from roads like East Cabbage

Patch Avenue to FM stations like Big Cabbage Radio are named after a veggie best known in other parts of the country for being doused in mayo and served as slaw.

The soil at 12430 East Drift Lane once birthed these giants. Rutabaga swelled up to 82 pounds and celery stalks weighed in at 63. The cantaloupe were as heavy as an adult husky and kale rivaled the weight of a 32-inch flat-screen television. But by far the most impressive of Scott Robb's crop was his giant cabbage, and one season, one head of it grew to 138 pounds, the largest ever to sprout from Earth. The gravity-defying produce at his "Colony Greenhouse" broke world records, which made him a Spartacus in the international arena of giant vegetables—yes, such a world exists. But when I pulled up to this storied soil, all I found was what looked like dirt. Scott's Eden had been abandoned.

Like Scott's current location, his origin story also isn't clear. According to some, he's a descendant of the Matanuska Colony. According to others, he moved from Michigan. One person told me both were true. Either way, everyone agreed he seemed to have some generational knowledge of the seasonal alignment of the stars and the moon and their astrological power to grow colossal food. This was fe-fi-fo-fum folklore, and finding the man at the center of *this* fable was going to be harder than I thought when my producer Walker and I first visited Palmer in 2022, during the eighty-sixth annual state fair. "He's one of Alaska's most elusive figures. Some people call him the 'Einstein of Produce,'" Steve Brown, a local grower and agriculture researcher, told me when I asked if he had a phone number, email, or address where I could find Scott. The two had competed together before. "I don't think he likes me very much, because I one time asked him what his secret for growing was. Everyone used to ask him that around here." He laughed before giving me the most recent number he had for him. "I can't make any promises it will work. When he retired a few years ago and closed his greenhouse, he sort of just vanished."

As the story goes, Scott's training for competition began every May, shortly after the last frost of the season, when the Sun danced along the horizon on an endless loop. Under this midnight star, he labored away around the clock, even sleeping in the field to protect his behemoth crops from behemoth moose. It was a twenty-four-hour-a-day obsession that culminated in early August when autumn arrives in Palmer and the curtain opens on the fair. While giant vegetables have been grown in Palmer since 1935, Scott was one of the first people to tap into the mythic nutrient-rich soil and sun in a way that caught the agricultural community's attention as Guinness world record after world record—five in total—were broken. "People started to say, what's in the soil up there in Alaska?" Steve recalled. Scott hung his hat up at the age of sixty-six, after more than two decades of victories in his field. Pictures from past competitions show a man with a thick brown mustache and glasses with tinted transitional lenses. He always wore a khaki-colored baseball cap. Scott looked like a most-wanted man, and every lead I had on him came to a dead end. The phone number Steve gave me was disconnected. My best bet was a boots-on-the-ground manhunt at the fair, but from what little I knew about Scott, he didn't seem like the type to return to the ring for nostalgia's sake.

I got a close look at other growers' prized fighters on display in the "Showcase of Giant Vegetables." A purple ribbon was attached to one cabbage's wilting, waxy leaf. Its green rosebud-layered face resembled the carnivorous plant from *Little Shop of Horrors*. It was impressive, and yet only a toddler compared to Scott's record-setting Audrey II. I took a few photos before making my way through the artery of the main hall, which was clogged by a mass of gawkers staring at other freakshow-worthy vegetables . . . what one person called the cabbage's "supporting cast." Deb Blaylock's onions weighed more than a pound. Theresa Philips's tomato plant reached 12 feet tall and her husband Thad's fennel was 7 feet long. Then there were the pumpkins, which sat on the hay-covered ground under hazy fluorescent lighting like orange Jabba the Hutts. The season's largest one, grown by Dale Marshall, weighed 2,147 pounds. "Out of curiosity, have you seen Scott Robb?" I asked one of the women monitoring the exhibit. "I don't think he's been around here in a few years," she said. A vendor at one of the farming booths near the showcase suggested I contact the fairground's main office, where all the information on past winners was allegedly stored.

"Thank you for calling the Alaska State Fair," a woman named Pam said when I called to introduced myself. "Have you spoken to Kathy Liska? She's the crops superintendent and has everyone's contacts. If you have a pen I'll give you her email. She's good at responding." I thanked her and hung up with a small measure of hope.

Hi Kathy!

Pam at the State Fair said you would be the best person to contact regarding information for past winning growers. I'm trying to track down a good number for Scott Robb and am hoping you may have one.

I look forward to hearing from you.

Warmly,
Jonathan

I clicked send on my email before searching for more clues in archived articles I had saved on my phone. In a rare interview Scott gave after breaking the world record for largest cabbage in 2012—the one that weighed 138 pounds—he credited his seeds and old-fashioned hard work. "You're not going to win the Kentucky Derby with a mule or a Shetland pony. If you don't have the right genetic material, you're never going to achieve that ultimate goal," he said, adding that growing in Alaska required "persistence, perseverance, and patience. There's no two ways around it." By the time I finished reading, Kathy had responded.

Hi Jonathan,

I have an agreement with Scott Robb not to give out his number, but can contact him if I know what sort of information you are seeking to see if he is open to communicating with you.

Regards,
Kathy

I explained I was looking to learn more about Robb's background and his rumored lineage from the original colony before thanking her for forwarding my request. My search had a pulse. A faint one, but a pulse nonetheless.

It turns out persistence, perseverance, and patience, along with the right genetic material, are also key to any worthwhile reporting pursuit. And maybe both journalism and agriculture share another common denominator: luck. When colonizers first arrived in 1935, the land was divided into 40-acre squares and, as is the case on chessboards and on Earth, some squares are better than others. Plots located closest to the mouth of the Matanuska River, like Scott's Colony Greenhouse, benefited from nutrient-rich floodplains, glacial silt, and volcanic ash. If some of the legends were true, Scott inherited his prized land and built his pioneering success.

"The early idea of establishing a colony may have been poorly executed, but I think we're now seeing there was merit. I mean, there was a reason why government scientists picked Palmer for the colony. They were just a few decades early," Steve told me during a follow-up call, and clarified that growing small crops of massive vegetables was a different business than farming hundreds of acres of small ones. "But what Scott did, and really what all the growers and farmers here have proven, is Alaska is fertile. Is it more fertile than the lower 48? I think it's fair to say the land is less spoiled," he said. "One farmer once told me—and I'm not being tongue in cheek here—his soil has been organic for the last 10,000 years at least. The point being not a single chemical ever touched his soil since the last ice sheet retreated. Now that's 'organic' like no one's ever seen before. The truth though is the state's soil historically has been terrible because it was mostly frozen." I asked him what impact, if any, climate change was having. "You hate to say climate change has been good, but in the farming world, it's been good. In the last few decades or so our soil has become better for farming. There's no doubt." Alaska's big thaw was happening as events like the state fair were attracting more international interest.

It's unclear how much longer Alaska's soil will remain unspoiled. The state's climate has warmed around 4 degrees since 1970, according to the federal government's national climate assessment. That's twice as fast as the rest of the world. It's no longer rare for summer highs to reach the 80s and even 90s, 30 degrees warmer than normal! Land whose cultivation was once a labor of love is now opening up to larger-scale operations focused on bigger yields of smaller vegetables. A new wave of pioneers are once again flooding the region, and this time the infrastructure is in place for rapid expansion and easier access to the state's natural resources. According to the Alaska Farmland Trust, from 2012 to 2017 the number of new farms in the state increased by more than 44 percent as states in the lower 48 lost farms at a rate of 7 percent over the same period. Believe it or not, Alaska is now the leading state in the nation for new farms.

———

Walker and I left the fairgrounds to meet Zoe Fuller and Nick Treinen, both in their late twenties, as they worked barefoot in their organic field, pulling carrots from the soil and placing them into plastic buckets. They were on a tight deadline. Zoe was expecting her first child any day. The couple were the youngest farmers I'd met in my years covering agricultural stories across the United States. They reminded me of a young version of Barb and Jim Kalbach . . . this generation's elusive small family farmer. They had both been in the field all day but looked fresh-faced and sun-kissed. Their golden complexions could have also come from the beta-carotene in the carrots they'd been snacking on. "Try one. They're like candy," Nick said as he handed me one straight from the ground. It tasted like it had been dipped in sugar, with bits of soil giving a Sour Patch Kids–texture. "This is the peak of abundance here right now in September. If you came here in November, December, January, February, March, April, it would be covered in snow. Even May, June, and July can be tricky," Nick said. The back end of the harvest can also be equally unpredictable.

"I'm remembering a couple of falls ago when we went to bed planning on harvesting carrots the next day only to wake up to four inches of snow on the ground, and it was still snowing," Nick said as Zoe laughed. "The ground was frozen an inch or two down and it was all-hands-on-deck trying to get the carrots before they froze." I asked Zoe if they were able to save the crop that year. "We sure did," she continued, laughing. "It's a super short growing season."

Both had been born in Alaska, and Zoe studied sustainable agriculture in college, a passion that brought her back to her childhood home where she began farming in 2019. She made a point of saying her family had no ties to the original colony—her mother had moved to Palmer for the National Parks Service when Zoe was a kid. Today, Zoe and Nick sell most of their produce at farmers markets. "Despite the challenges that still exist, it feels possible and important to grow food here," she said. "Ninety-eight percent of our food [in Alaska] currently comes from the lower 48. I think localizing our food supply is important as a means to not rely on industrial agriculture and as a means to steward the land while providing communities with locally grown food," she said. "While warmer temperatures will help make for a longer growing season, we're also aware of how this pattern has played out in other states. At what point does it get so warm we run into issues with accessible water? We've already experienced droughts and record-breaking wildfires." A study by the University of Alaska found the state's warming weather contributed to a 17 percent increase in lightning storms in the last three decades. In the same time period, the amount of land destroyed by fire has doubled. Scientists say the two are connected.

Palmer in the year 2022 had the small-town energy and excitement I imagine 1960s Dexter, Iowa, had. I felt a sense of hope for a new way of farming. But in another familiar pattern, factory farms and developers looking to build apartments and homes were starting to circle. And it wasn't just happening in Palmer. Six hours north, near Fairbanks, the state had begun a new experiment in one of Earth's most critical ecosystems for helping fight climate change. If all went according to plan, part of Alaska's pristine boreal forest would soon join America's agricultural fray.

In the end, I never found the Cabbage King Scott Robb, but months after the 2022 Alaska State Fair closed its doors, just as Nick and Zoe and their newborn were getting ready to hibernate for the winter, he contacted me. The message from Palmer's "Einstein of Produce" unceremoniously came in email form.

Hello Jonathan,

I'm flattered but not interested. Happy holidays to you and yours.

It was the first and last time I heard from what history will consider one of the original pioneers of Alaska's modern landscape. I wondered if the state's agricultural future he helped plant could avoid the cycle of nurtured abundance followed by corporate greed unfolding in the rest of the country.

Checkmate

Minnesota's winter twilight is the perfect chalkboard for the luminous clouds that billow from the smokestack at the Blandin paper mill—the one that produces most of the glossy pages found in America's magazines. If the wind is blowing northeast, you can follow the white streaks, illuminated by the last rays of the retreating sun, to a small nondescript building that sits in the mill's shadow off Route 2. But you really have to be on the lookout, because many buildings in the town of Grand Rapids, 100 miles south of the Canadian border, look like bunkers built to escape the minus 30-degree Fahrenheit February lows. It also didn't help that the blue-sided, one-story bungalow I was looking for had no visible signage. I guess there really wasn't a need. Most visitors don't walk through the front door.

The phone lines that connect 1007 Fourth Street to the rest of the state are monitored twenty-four hours a day, seven days a week, including all holidays and, in the nonprofit's short history, one pandemic. In the first few months of Covid-19—when it was unclear how the virus spread and how deadly it was if contracted—the small team of friendly voices inside the first-of-its-kind call center created their own "bubble" and didn't leave for forty-two days. They slept on blow-up mattresses in the conference room under a sign that read "If the world 'says give up,' hope whispers 'try it once more,'" and some "bathed" out of the bathroom sink in between answering calls and heating microwave dinners or ordering in 12-inch pepperoni pizzas. They were on-call all day and night because, deadly virus or not, American farmers—their clients—couldn't afford to take a day off.

Thirty thousand calls come into 1007 Fourth Street every single year. That's more than eighty-two calls every single day. The lines light up at all hours, but the traffic picks up at night along with the desperation in the voices on the other end. "Some of the farmers call in the middle of the night. Two in the morning on their tractor, trying to get the tilling done, because they know seeding has to happen—'Can you just talk to me? Can you help keep me awake?' They've been up for two days straight and can't afford to take a break," Cre Larson said as she showed me around the offices of the Minnesota Farm and Rural Helpline, which she directs. Cre had a soft voice with a slightly drawn-out accent made of rounded vowels and sharpened "*t*"s arranged in a singsong composed over fifty-five years living in Minnesota. The toll-free number was created in 2018 to address the alarming rate of farmer suicides in the region. "Just in the last two weeks, two farmers took their lives," she said before a long pause.

What's happening in the state is part of a growing epidemic sweeping rural communities across the nation, turning farming wives into widows clenching suicide notes. "My dearest love, I have torment in my head," one of them read. According to the Centers for Disease Control and Prevention, suicides among farmers have increased by 40 percent in the last two decades to a rate that is more than double that of war veterans. A report by the Midwest Center for Investigative Reporting found 450 farmers across nine Midwestern states including Minnesota and Iowa took their lives between 2014 and 2018. The number is believed to be even higher because many farmers disguise their suicides to look like accidents so their families can collect life insurance. Providing for the family used to be about blood, sweat, and tears, but now farmers, desperate and depressed, are falling on their own swords. "What we hear is a lot of fear. Fear of failing and fear of letting down the family. There's a lot of fear and shame," Cre told me as one of her crisis operators, Crystal Terveer, finished a call with a farmer. It was her tenth that day, and she wasn't even halfway through her shift.

"Shame is what holds many farmers back from reaching out to get help in the first place," Crystal told me during a brief lull. "Many calls come from their wives. I once received a call from a woman worried about her aging husband. He wasn't accepting any help. They didn't have kids that wanted to take over the farm and he refused to give it up. He had recently broken his hip but still got on the tractor, and she was very concerned about him so she called for emotional support because he refused to accept the change. These were people both in their seventies so change is understandably hard," she said. "Based on the conversations I have, it just seems to be getting worse and worse, and there's very little reprieve. Bills are due. The mortgage is behind. The price of gas and seeds and labor is up, but then the crops that all of this investment is for have been bad. During the last summer, the drought was so bad their crops just failed and the prices for the crops they did harvest weren't good enough to cover the costs of farming. It seems to be getting worse every year," she said before the red light on her phone began flashing. Another call.

The traditional solution for many small farmers was often squeezing more work into every hour of every day to maximize output, but elbow grease is no longer enough. While farming used to be a game of strategy, thanks to that rigged competition with factory farms and the unpredictability of weather fueled by human-caused climate change, there's only so much a family farmer can now control. And for those out of moves, the good old chessboard becomes a dizzying game of roulette, where many of those around the spinning table can't even place the kinds of bets that lead to the kinds of winnings to break out of the red.

"That's when things begin to get really desperate," Cre told me. "They'll start selling off their cattle and their equipment and even some of their land. These are farmers who have, in many cases, inherited their land from their parents and grandparents, and they feel like they've somehow let their families down. When fear turns into desperation,

I don't know how to explain it—it's like the body just genetically changes. That desperation makes people do things they would never do, that go against our basic instincts to survive and live." The helpline has become a lifeline for the pawns who struggle to imagine a life after the game. I pictured that farmer Cre told me about—the one out in the field at two in the morning—feeding seeds into rows of slots in the soil, then cranking his tractor into gear, hoping to hit a jackpot in a cash crop casino where the house always wins. Walking away isn't an option for those addicted to the land.

The call center is only the first line of defense. Cre and her team ask everyone who phones in if they'd be open to free face-to-face chats. "We'll go anywhere. Coffee shops. The caller's living room. Sometimes we'll just walk around town. Some of the best interventions I've done are just walking. Leak that valve. Walk in the street, fields, their yard. Anywhere. We just need to find one thread, and then we can use it to pull them back from their desperation and build a safety plan and a road to recovery. We got a call today from a family member who thanked us for saving their loved one's life. She said we had met with her family member, and after that he had changed. He started to see ways out of his crisis. That's why I don't sleep. That's what pushes me to keep doing what I'm doing," she said.

But not every thread can be found, and Cre has lost farmers. "His wife was the first one to contact us. Her husband was eighty years old. It's a scenario we see more often than not. They had been married for sixty years. He had just retired and sold off his farm and he was really struggling because his children didn't want to carry on the family business. I ended up meeting with him and we spoke for a while. He was concerned about his failing health and being a burden on his wife. They had no family around to look after them. We discussed options and I thought we had made progress." Cre later found out he took his life with a shotgun. "These were supposed to be the most peaceful years of his life. Your heart just breaks. It's just another reminder why we do what we do," Cre said.

Cre's support staff are trained to listen to callers and give solutions to surmountable problems. Oftentimes, farmers are unaware of state or federal programs that can provide financial relief during a bad season. Many calls come from people who just need to release the tension that's been bottled up inside. Leak that valve. The worst calls though, like the late-night one from a farmer gripping a shotgun and a glass of whiskey, are patched through to psychologists like Ted Matthews.

"Farming is, for many, a calling, and there's a deep psychological toll—an emptiness—when you can no longer survive off your land, and unfortunately it's a very difficult calling even for the best of them. It's overwhelming how many things can go wrong, and often do these days in farming. There's not a minute where a farmer doesn't feel stressed. There's always something going on that they're going to have to deal with or simply can't deal with," Ted told me from his wood-veneer-paneled office three hours south of Grand Rapids in Hutchinson, Minnesota. He was a trombone of a man who emanated a smooth and calming tone that rarely slid to a high octave.

Ted was once the director of FEMA's mental health strategy for responding to disasters before working with the Minnesota Department of Agriculture to develop the state-funded helpline. Both roles shared striking similarities. "What's happening in rural America is a slow-moving but very destructive disaster. Think about this: A typical farm fifty years ago, four hundred acres was plenty to get by. Now four hundred acres is considered a hobby farm. It's hard to find anything in these small towns anymore. Downtowns are just dying. If you were to be dropped into one of these communities and not know where you were, you'd look at these abandoned buildings and think maybe a hurricane or flood swept in. They all look the same," he said. "Now imagine what it takes to live in a community like this where there is no support system. Some can hold on. Maybe they have more savings or more land or a bigger family all helping. If you don't have a network, it's tough to get by, which doesn't make it easier to let go and move on. What most people don't understand about farmers is, this is a way of life for them. It is not an occupation. It's not like, 'Well, if they don't do this they can do something else.' This is who they are.

If you're a third or fourth generation farmer and you lose your farm, that's all on you, and the fact that it wasn't your fault, that the weather or tanking prices are to really blame, doesn't help ease the pain at all." As Ted put it, some farmers can't or won't envision an Earthly afterlife. Checkmate in the game is checkmate in life.

While it's difficult to measure the overall success of the helpline, Ted believed the service provided an outlet and a sense of community that was otherwise being lost, but he admitted it's not enough. "The government needs to recognize this crisis and take action." Existing anti-trust law currently allows Washington to stop the kinds of big mergers that give factory farms and the supermarket chains they supply control over the industry, but those laws have been largely unenforced. For now, focusing on mental health may be the only way for the pawns in this rigged game to weather the crisis. "The writing is on the wall if the current farming structure doesn't change. My fear is corporate farms are going to take over, and this will have many implications to livelihoods and the land," Ted said.

The Minnesota Farm and Rural Hotline is one of a handful established in states across the country in what is a new chapter in modern American agriculture. In places like Alaska, meanwhile, an old chapter is being reworked.

Virgin Land Grab

The Sun's surface, the photosphere, is a sea of explosive plasma that sprays clouds of electrically charged particles into the Milky Way, where they're magnetically pulled to our planet's south and north poles. When these solar protons and electrons collide with Earth's atmospheric gasses, they spark billions of tiny flashes in a galactic orgasm that showers the sky in shimmering shades of neon green. The tint of cosmic intercourse. Alaska's aurora borealis is also the color of opportunity, and the state's virgin interior—where they seem to flicker most these days—is now open for business.

For a limited time, you can buy your very own piece of the last frontier—a 32-acre parcel off "Moe's Road" starts at $15,500, and the 183-acre parcel down the street on less desirable land is only $39,400 (around $200 an acre!). The only catch: You need to farm. Or at least have the intention of farming—it's not exactly clear in this buy-now-figure-it-out-later land grab in one of the most environmentally significant habitats on Earth.

The boreal forest, also known as the taiga, is the world's largest terrestrial ecosystem. It covers more than 1.3 billion acres and spans eight countries, including Canada, China, Finland, Japan, Norway, Sweden, Russia, and the United States. This biome of mainly pine, spruce, fir, and birch covers about 11 percent of all land on Earth and is the planet's largest carbon storehouse, sequestering around 208 billion tons of it, or about 22 percent of the world's total. The forest is often called a "carbon sink" because of how the habitat's frozen soil traps the greenhouse gas in its icy subterranean drain. But as Alaska's temperature rises faster than anywhere else in America—and most everywhere else in the world—this permafrost is thawing, releasing carbon back into the atmosphere and attracting the kind of human development that's helping unearth even more.

Walker and I met the group of government "handlers"—a friendly team with the state's Department of Natural Resources—in a gas station parking lot where we shook hands, talked about the weather, and silently assessed if the tour ahead would serve either of our interests. Our handlers were proud of their well-intentioned, but half-baked, scheme to bring farming to the boreal, and I immediately knew they would never want to talk to me again after the piece we were working on aired. The Nenana-Totchaket Agriculture Project, or Nen-Tot, as locals called it, took decades to develop, and we were told would be rolled out over the next thirty years in phases. Every year, thousands of state-owned acres, divided into parcels, would be auctioned off to the highest bidders, who would then clear them and cultivate them for farming. A total of 140,000 acres were up for grabs. There was no electricity, running water, or sanitation. This community, whatever it would be, was starting from scratch. We arrived weeks before the project officially broke ground to survey the land set aside for this twenty-first-century Matanuska Valley Colony. As we convoyed down a freshly paved road carved into the heart of the forest, Walker and I slowed to photograph a lynx slinking through the tall grass. The notoriously shy animal looked confused to see humans on hunting grounds where, up until now, he was the apex predator. That would no doubt change, I thought.

"This is a project of necessity. Most of our groceries are currently imported from the lower 48. Around 95 percent. Our food is expensive, and our supply is insecure, which became clear during the pandemic when supply chain issues slowed everything down, including food delivery," project leader Erik Johnson told me as we walked through a plot of rugged forest that would be part of the first round of the auction in a few weeks' time. Erik was a grizzly of a man in stature (he was well over 6 feet tall, with a long brown beard and thick hands) but a teddy bear in demeanor as he talked excitedly about his vision for Alaska's agricultural future. "I know the challenges that the state has run into in the past. But everything is working in our favor this time around. So much has changed."

Erik knows those "challenges" because his grandparents were among the original group of colonists the federal government brought to Palmer in 1935. "My grandfather's farm in Minnesota was hit by three years of bad luck. First it was destroyed by a windstorm, then there was a devastating frost, and the last straw was a fire that burned it to the ground. He was on welfare when the government asked him if he'd like to join the new colony," he told me. And so Erik's grandparents packed up their things and journeyed to the great unknown, where they started a dairy farm. They were one of the few to not only survive the "before," but start a family that continues to grow today. Both Erik and his parents still live in Palmer. "Palmer was Alaska's first big agriculture project, and lessons were learned, but in the end I think the biggest problem was just that it was ahead of its time." Alaska has an agricultural history of "ahead of its time," including the "Barley Project" in northern Delta Junction. In the early 1980s, state land was broken into plots and sold to farmers for the purpose of growing barley. The plan was to transport the grain by rail to Alaska's coast and ship it to Asia. But like Palmer, the barley project was largely considered a failure, due in part to the region's arctic weather, which made growing nearly impossible. "The weather is different now. Both Palmer and Delta Junction are starting to thrive," Erik said. "I see climate change here in Alaska as an opportunity to bring in more crops, to develop more land in a way we never could before."

I liked Erik. He spoke in the unguarded language of a believer. If he was fed specific lines to tell me—as many officials on similar government tours are—I couldn't tell. It also helped that reality supported what he was saying. Alaska imports more of its food than any other state in the country. A walk through any Alaska supermarket, where fresh produce can be sparse and a family-sized box of cereal can cost more than $10, is all it takes to understand that America's forty-ninth state clearly needs an independent supply of food. "But couldn't you look anywhere else to develop the land than here in the boreal forest?" I asked.

"We're only looking to auction off a total of 140,000 acres right now in a state with more than 300 million." That's nothing, was his point. But it was also clear if all went well—and more and more the odds are in Alaska's favor—this exploratory mission would pave the way for more roads through more of the world's carbon sink, turning the trees and tall grass of an ecosystem where the lynx roams free into tilled rows of crops.

Erik took us farther down the road to a football field–sized piece of land that had been cleared down to the soil and topped off with a thin layer of gravel. It was the foundation for a town center. "We're trying to think forty, fifty years from now and what will be needed. Services like a school and firehouse. We can't expect people to farm this land and then need to drive forty-five minutes for gas and food." I was standing on America's newest main street on a pie of pristine land large enough and already primed for many more.

———

Indigenous tribes don't consider land a commodity. For them, Earth—like Air, Water, and Fire—is not something to own, but to live with and protect. Unfortunately, most other cultures are the polar opposite, and the ones that came into direct contact with natives always had more lethal weapons, which ultimately gave them the power to call the shots. In 1971, twelve years after Alaska became an embroidered star on the American flag and nearly a century after the U.S. stripped all native tribes of their land, Congress passed the Alaska Native Claims Settlement Act, which gave back some 40 million acres to tribes, including about 24 percent of the state's boreal forest, with the remaining forest land sliced up like a pizza and handed over to different interests. The federal government controls about 51 percent, and universities and state and local governments own another 25 percent, according to the Boreal Conservation Program. Most of this land is unspoiled and still available for native tribes to use for subsistence living as a kind of colonization consolation prize. Less than 1 percent is controlled by private landowners, where native tribes are usually not welcomed. While this private land accounts for only a fraction of the whole, this slice tends to be loaded with the most toppings. We're talking about land that often has the best irrigation, abundant wildlife, and access to rivers and roadways—after all, why else would these private owners be incentivized to pay? Initiatives like the state's Nen-Tot Project are slowly transferring more and more of this high-value soil to individuals and taking it away from its original guardians, who are already struggling to make the most of their marginalized slice.

Paper-light pieces of scraped moose hide flurried around the group of women when Walker and I arrived in the afternoon on their native soil next to the Nenana Municipal Airport. In Indigenous Nenana Athabascan culture, Earth and all her animals have a strong and sensitive spirit and must be treated with great respect—even the bullish moose. Not a single part of the animal is wasted after the hunt, and the complex process of refining raw hide into pliable leather is a labor born from equal parts necessity and reverence. First there's skinning, then "wet scraping" off the flesh and hair, followed by stretching. Next comes "dry scraping," smoking, and tanning. The final steps include another round of stretching followed by a second optional round of smoking for added water protection. The transformation can take days, even weeks. The result is a soft hide that can be used to make everything from clothing and moccasins to bags and blankets.

The women were in the dry-scraping stage of their heritage leather-making process when we introduced ourselves. Several hides were stretched on 10-foot-long wooden frames. The breeze off the Tanana River carried with it the smoke from a small fire where fresh-caught fish were being cooked for lunch. A man was chopping logs to help feed the flames. "This land, this is our grocery store. We grew up learning how to live off this land," Native Eva Dawn Burke told me as she showed me around her "fish camp," a compound inhabited every summer and early fall by natives and used to catch salmon and grow grains for sustenance during the winter. Eva was the leader of this compound. Her dark brown hair was braided into two long pigtails, which dangled behind her ears under a black baseball cap with the words "You Are on Native Land" stitched in white. She wore a matching black T-shirt that simply read "NENANA land." She was thirty-nine years old, but looked younger.

While Eva's fish camp was built on the part of the pie owned and supported by the Nenana tribe, most of their moose and fish came from other slices, including the land for the state's new farming project just ten miles upstream. "That's land we'll no longer have access to. And what kind of impact will farm runoff and other pollutants from development have on the health of the river? Our culture is entwined with the environment," Eva said. And that culture was quickly vanishing from Alaska's landscape. While several tribal members spent all summer living in the fish camp, including Eva and her family, most only visited during the day as part of a tribe-funded program to preserve culture through education. "It's crazy to think that when I was a kid, fish camp wasn't an after-school activity. It was the only life I knew, and dozens and dozens of families participated. I grew up spending my summers at fish camp and then my winters on the trapline," Eva told me. The "trapline" is a winter camp used by the Nenana and other Indigenous groups for hunting. "I was homeschooled and learned about my tribe's traditions from my grandmother, who lived with us. A lot of my friends at the time were leaving the camps to become 'cityfied,' but my family continued to raise me out on the land, and everything that we earned came from the land. Like, I didn't get my first pair of shoes until I was in second grade or something," she said before offering us a traditional bowl of akutaq, a savory/sweet mixture of cooked white fish, berries, seal fat, and ice whipped together and consumed as a protein-packed snack while out on the hunt. "Agriculture is probably something we need to get into as a state, but what does it look like? It doesn't look like this," she said, referring to the state's new auction. "Have you read the terms of the sale? Basically, anyone can buy the land and do whatever they want with it."

Those terms were sprinkled over seventy-two pages in the "2022 Alaska State Agricultural Land Offering" brochure, alongside maps of each of the twenty-four parcels up for grabs. And Eva was right—nowhere in the fine print were buyers required to actually cultivate the land. Instead, they only had to commit to clearing at least 25 percent of the parcels within five years and keeping it in "farmable" condition, meaning free of trees. The auction was open to anyone from anywhere in the world, including businesses and corporations, and there was nothing stopping a single person, business, or corporation from buying all twenty-four parcels. "When I hear that, I think many things. Land grab, for starters, but then, what will that land be used for? What's to stop big factory farms from moving in or massive housing developments from being built?" asked Eva. "This is the boreal forest we're talking about. It's sacred land."

Eva's concerns were valid, and not eased by what Erik told me when I had spoken with him during our tour. "We want real farmers. We want to provide opportunities. We can't tell them exactly what to do with those opportunities," he said. It all sounded like a take-me-at-my-word "handshake deal," and those rarely end well.

Turning the (Dinner) Tables

The porous aluminum siding in Dexter, Iowa, had calcified over years of baking in summer heat and cracking under the town's winter frost. The once-solid coat of strawberry-red paint had taken on a muted and blotchy pallor. The exoskeleton of Jim Kalbach's workshop had been built and added on to during good times in a game that was once marked by gains before becoming one of survival. The unplugged Coca-Cola machine that stood in the main entrance was a marker of better days, a dusty reminder of when farmhands only paid 50 cents a pop. "I can't remember when that was, but time has changed," Jim told me after his wife, Barb, introduced us.

Jim had a firm handshake and slanted eyes that mirrored mine. The lines around *his* were deep from decades spent in the field, squinting in the afternoon sun. His oil-stained overalls smelled of sweet benzene, and he had grease under his fingernails. He never wore his wedding ring around the machinery, but an indent in the skin marked its resting place of forty-nine years. "When we first married, we used to do it all. We had livestock and grew corn and beans. We fed our cattle and hogs with what we grew here, just like our parents before us," he said as he tightened bolts on his faded green John Deere tractor. "Now there ain't nobody left. The big guys come in and rip everything out. Timber. Pasture. Vegetation. All of it is gone now. The soil here is terrible. Horrible. You gotta pump it with chemicals to get it to grow anything these days."

As America's soil has increasingly been treated like dirt—and as habitat and climate change have made it harder for modern methods to restore this dirt back to soil—yields have dwindled in a familiar pattern of crop failure last seen a century ago. According to a study by Kansas State University, the 2022–2023 wheat growing season was the driest in 128 years. Its author, Dr. Xiaomao Lin, characterized that as even drier than during the Dust Bowl. And it's not just wheat. Rising temperatures, shrinking rainfall, and mismanaged land are having a domino effect on nearly every American crop.

In 2022, corn yields were the lowest in ten years according to analysts at Rabobank, which collects data about commodity markets. In Texas, the USDA reported cotton farmers abandoned 70 percent of their harvest because it was so paltry. And in California, the rice harvest shrunk by 50 percent. A study by the United Nations–backed Global Commission on Adaptation estimates that worldwide crop yields in traditional agricultural land could decline up to 30 percent by 2050 as the planet warms and fuels severe droughts and floods. "When you're a small farmer, you don't have the kind of land . . . the kind of acres to absorb major hits like that," Jim said. "You just can't survive."

For the Kalbachs, survival has meant abandoning livestock altogether and focusing on the only thing that pays the bills: corn and soybeans. The traditions passed on to them by their families had, over time, been lost—but the hardest part of it all was where the fruits of their carefully grown, grafted, and now pruned-back family tree were going: the very factory farms that were hollowing out their neighborhood and the rest of rural America. Unable to compete with the "big guys," they were forced to sleep with them. For more than a decade, the Kalbachs have sold their crops as feed for livestock . . . on factory farms! It made me think about that statistic from the USDA that showed while America's small farms represented about 80 percent of all agricultural land in the U.S., they were only responsible for

20 percent of production. While that was bad enough, I now realized what actually wound up in supermarkets was even less, considering some small farms like the Kalbach's only existed to support larger operations. Their produce never made it to human mouths.

"Once you give up the land and hand it over to the big guys, you ain't never getting it back. There's not gonna be any small farms. It's all gonna be the big guys who don't know nothing about this land." It was a bleak assessment, and as I left Dexter at dusk that evening, with the oranges and purples of the sun's setting rays flickering between the gaps in the wooden slats of rotted barns on Dexter's fallow land, I struggled to imagine a solution to restore balance to this chessboard. With small farms representing less than 20 percent of what you can find on the shelves at your local grocer, consumers have lost the kind of purchasing power to help turn things around. Was I driving through the last few moves before a checkmate? Yeah, according to Jim. But perhaps there's an opportunity to reset the board.

While climate and habitat change threaten these ecosystems, there are potential opportunities if we learn from our past mistakes. Increasingly, places like Palmer, Alaska, could become vital growing grounds—providing food to people and relief to soil that's been treated like dirt in overworked fields like Dexter. The key, it would seem, is to find sustainable Earth—not the boreal forest—and keep it in the hands of stewards. And as I've learned in the field, those stewards are often the small guys.

As for the Nen-Tot land auction in the boreal forest . . . it was the kind of success project leader Erik Johnson hoped for. Twenty-four parcels were sold to fifteen bidders, for a grand total of just $1 million for roughly 2,000 acres. It was the first of many more auctions to come. Of the winners, twelve had mailing addresses in Alaska, though Erik told me that did not mean they actually lived in the state. "They could be from other parts of the country or somewhere else in the world." Three of the bidders had addresses in the lower 48 and collectively owned about 30 percent of the pie. Alaska officials did not release the names or companies associated with the winning bids, but I do know one.

"We couldn't believe it," Eva Dawn Burke told me over the phone. "We could only afford 25 acres, which is so small, but it's something. We finally now have legal rights to land that we always considered ours." Eva, and the Nenana tribe she belongs to, had ripped a small page from corporate America's playbook by rewriting the rules of the game. And while 25 acres was really only a crumb of the pie, it was the beginning of more to come. "We're raising money. We're going to buy additional land. If Alaska becomes the future for food, then I think we have an opportunity to create a new way forward where we can sustain people and nature. We don't have to sacrifice one for the other," she said. "There's still time to protect our planet and to restore balance. We don't have to accept things the way they are or the way they've been done for decades and centuries. Those ways haven't worked out well, have they? If we don't change our behavior, we'll spiral out of control and only have ourselves to blame."

Dust Bowl Redux

On May 13, 2022—four months after I visited Barb and Jim in Dexter, Iowa—an unusually powerful storm packing hurricane-force brawn hammered neighboring Nebraska. Millions of tons of loose dirt from drought-stricken, overtilled farmland revealed the weather system's biceps. The swollen muscle of silt, sand, and clay reached miles into the sky and cut visibility down to only a few inches. People trapped in the middle of this atmospheric and earthen arm couldn't see the destruction, but they heard it as the fine particles bruised everything in their path. Local officials declared an emergency and shut down roads as the storm lumbered eastward for 240 miles, pummeling parts of Iowa and South Dakota for more than five hours. Local meteorologists referred to this monster as a "haboob," a term commonly used in places

like Saudi Arabia, the Sahara, and Sudan. A television weatherman mar-
veled at the phenomenon and called it a "once-in-a-lifetime event." But
those with longer memories recognized the circular lifeline of history
repeating itself. The last time a "haboob" swept through this region,
America's breadbasket was called a dust bowl. The beast Dr. Hugh
Bennett helped slaughter nearly one century ago had a new pulse. It
was faint. But a pulse nonetheless.

Epilogue:
Kamikaze Iguanas

It was January 10, 2010, and South Florida was colder than parts of Alaska for the first time anyone could remember. A waning crescent moon lit my search along with the flashlight from my iPhone, which said it was 35 degrees outside, with the windchill making it feel below freezing. "I've got an urban legend for you to chase. The conditions are perfect," my assignment editor said before launching me on this fool's errand. I've been on many in my time, but this one, on a golf course in the city of Hollywood, felt particularly fucking foolish. As I made my way over fairways and through sand traps into the rough, I heard the sound of tree branches breaking in the wind and the thud of stray golf balls landing around me—except it was two in the morning and no one, even in their twisted mind, would golf this early.

I was twenty-six years old and working for WPLG, the local ABC affiliate, as what the industry would soon call an "MMJ," or multimedia journalist. It was code for cheap labor that can shoot, edit, and report their own stories. I loved this scrappy style of reporting and to this day am grateful to my boss, Bill Pohovey, who gave me the job even after I screwed up my application when I submitted a story pitch about a nature photographer, but accidentally uploaded the wrong images (these were the early years of online dating and the land mines were plenty). But I digress. My small footprint as an MMJ gave me fly-on-the-wall status and access to some of the most intimate moments of my career. I also quickly learned Bill and my other bosses often forgot about me while on assignments, which allowed me looser deadlines and more time to dive deeper into a story. Which is what I did on that morning as I first searched for the perfect spot, set my camera on a tripod, and pressed record on my subject.

Iguanas have no business being in South Florida. That's why they're called an invasive species—which, if we're being technical, we humans are as well. And *we* are how the alien reptile, with its mohawk set of spikes, stout nose, and marbled blue/green scales first made its way from Central and South America to the U.S. to begin with. Starting around 1960, shortly after the first *Godzilla* debuted in theaters, man got the crazy idea that these mini Godzillas—along with pythons, tegus, lionfish, tree frogs, Nile monitors, peacocks, and Rhesus macaques—would make perfect pets. Soon, pet stores started selling the iguanas next to cats and dogs and sad-looking Siamese fighting fish. As demand increased, so too did the supply. In the late sixties, flimsy greenhouses for exotic reptiles started opening around South Florida with baby iguanas stored on shelves in small plastic Chinese takeout containers with holes poked in the lids. These hatchlings would then end up in those pet stores and eventually in little Johnny's bedside terrarium before Mommy and Daddy got worried about how "cute little Godzilla" was actually getting fucking big, and released him into the wilderness.

Within less than a decade, these pint-sized mohawked monsters began appearing in trees on Ocean Drive on Miami Beach, and sunbathing on rocks along Biscayne Bay in Brickell. South Florida's wild iguana population grew slowly and for a while it was unclear if they would survive much less thrive in this foreign ecosystem. But then a hurricane called Andrew sped this ecological case study up when it slammed the region and swept all those greenhouse Godzillas in all those plastic *Mayflowers* away like Dorothys and Totos. The new wild world the survivors landed in proved a perfect petri dish for rapid colonization. In half a century, Florida's invasive iguana population exploded from zero to an estimated four million by 2010, impacting endangered plants and animals and even damaging infrastructure, like when a hungry iguana knocked out power to thousands after chewing through wires.

South Florida, thanks to a climate and habitat similar to the iguana's native one, turned out to be a pretty cushy second ecosystem, except for one small issue—the occasional cold snap. When temperatures dipped below 50 degrees Fahrenheit the iguana's blood, along with its motor functions, slowed down. Anything below 40 and iguanas went into a full-on state of hibernation. Since many lived in trees, it *could* actually rain lizards if the conditions were right. At least, that was the "urban legend" that spread after Floridians emerged from a cold snap a decade earlier and found iguanas belly-up on the ground. "Most of these cold-stunned reptiles survived once the weather warmed back up. I knew a gentleman who was collecting them on the street and just throwing them in the back of his station wagon; all of a sudden as he's driving down US 1, these things are coming alive and crawling on his back and almost caused a wreck," Ron Magill with Zoo Miami told me. Why I never asked what the man was planning to do with those iguanas is beyond me.

Which brings me back to that Hollywood, Florida, golf course and twenty-six-year-old me, drinking probably one of the first of what would be a lifetime of sugar-free Red Bulls, while cursing my frozen existence in paradise. And I'm not kidding. A small ravine of sitting water had actually formed a thin layer of ice. My camera had been rolling for thirty minutes and my fingers were white as the small arteries in their tips narrowed in the cold. My lens was focused on a hibernating iguana precariously resting atop a tree branch with its long tail dangling in the breeze. It looked like it would be rocked off-balance at any moment and yanked by gravity. It was one of at least ten I could see from the patch of soil I was recording from. And it didn't take long to find the source of what sounded like stray golf balls. Several iguanas fell around me as the wind shook the trees free of their scaled squatters. Holy fuck, it was actually raining iguanas! And soon the same recipe of wind, cold, and gravity triggered my iguana's kamikaze plummet. All of it was captured on film.

My report, once written and edited, aired on the six p.m. news that same day before being uploaded online, where it received more than a million views in less than twenty-four hours and nearly crashed WPLG's website. Diane Sawyer also featured it on the *World News* and CNN aired it around the globe. Thousands of people emailed, commented, called in, and wrote about their own raining lizard experience. Florida, home of extreme weather, had never weathered a storm quite like this one before. And because climate and habitat change were little-talked-about crises at the time, my video was packaged and presented like a segment on *Ripley's Believe It or Not*. But make no mistake about it, this was an entirely man-made phenomenon. And while at that point it had been at least a decade since anyone could remember seeing frozen iguanas belly-up on the ground, in the years that followed, raining lizards became a near-annual occurrence, with Jimmy Kimmel still to this day airing the same grainy video I shot from that Hollywood golf course.

The kamikaze iguana, to me, has become both the poster animal for ecosystem collapse and a mascot for how we can take back control of the kamikaze mission we humans are on. You see, initially it was believed Florida's increasingly extreme weather would kill off the iguana population as cold snaps linked to climate change became more common. But research conducted by Washington University in 2020 found South Florida's iguanas, incredibly, were able to withstand lower temperatures than they could just four years earlier, and "endure climatic conditions that exceed their physiological limits." The paper, published in the peer-reviewed journal *Biology Letters*, found that most of the reptiles were as active in colder weather as they were in warmer weather. "They could now move at much colder temperatures than they did before," said lead researcher James Stroud. Some scientists believed, given the quick transformation, iguanas had adapted to their environment, similar to how people who move from the ocean to the mountains acclimate to the higher altitude over time. Today, the once-frozen iguana now struts around all cocky when bad weather hits their home.

Now humans are a very different species—this I understand—but if a lizard brain could save the iguana from our planet's changing environment, then I've got to believe our brains can save us, too. No other species on this planet is capable of reengineering a smarter way forward than we are—one that protects our climate and habitat, stabilizes our radical elements, reduces the impacts of environmental disasters, and lowers the threat of ecosystem collapse. We've already proven we've got some of the iguana's adaptive spirit in us.

Today, 30 percent of all energy produced worldwide is classified as renewable—including wind, solar, hydro, and geothermal—according to the World Economic Forum. In the United States, renewables are the fastest-growing energy source, increasing 42 percent from 2010 to 2020, according to the Center for Climate and Energy Solutions. Texas—the oil capital of the star-spangled banner—is now also the nation's largest producer of wind-powered electricity. The Lone Star state's comptroller's office says more than 28 percent of all power comes from wind alone. In Arizona, the state's largest electricity provider said it would be entirely renewable by 2050! Arizona is also leading the nation on another front that's worth noting and celebrating. In 2023, Governor Katie Hobbs made the unpopular but critical decision to pause development in communities that rely on groundwater, after a report from the Arizona Department of Water Resources showed the state's booming population would outgrow its water supply if action wasn't taken. Elsewhere, here and abroad, there are also large-scale commercial projects for carbon sequestering that, if successful, could reverse the warming trend we're currently on and even bring CO_2 levels back to preindustrial rates, according to some models. But as you now know, even if true, this could take a century, if not longer, thanks to a delayed feedback loop. For now, we're stuck with shit weather. But the good news is that climate is only one, slow-moving, piece of the puzzle.

Our habitat is much more resilient and bounces back faster, like what happened when overworked land was returned to its natural state during the Dust Bowl, or when beavers were reintroduced to a dried-up riverbed in Coalville, Utah. Stronger habitats can better withstand our extreme weather. Unfortunately, beavers won't be enough. We must also reimagine where and how we farm and ranch our land—and where and how we build our homes and towns. Alaska's boreal forest, Florida's water-surged sand, and California's drought-stricken forests may not make the most sense. We must invest in newer, better infrastructure that works with nature and won't crumble when the next storm hits. We must prioritize projects to collect rainwater and ethically distribute it. We must create an insurance industry that discourages risky new development, rewards safe development, and reasonably supports old development. We must invest in science that not only monitors our planet's health, but offers measures to avert disaster . . . and then actually listen to that guidance. We must continue to develop technology that can better forecast and track storm systems and issue lifesaving warnings. We must protect our Indigenous tribes—our original stewards—and America's poor, who always seem to pick up the tab for the wealthy.

We manifested this destiny and can manifest a newer, more sustainable one. I'm not preaching socialism or what some will call communism. In fact, the path we're currently on means we'll all likely *share* the pain. We must change to protect our independence. As Dr. Bryant put it, we must find and create spaces where we don't have to fight, defend, anticipate, and perform. We must find the space to breathe, be, create, and bloom. We must. And if the kamikaze iguana can teach us anything, we can.

Before it's gone.

Acknowledgments

My love for our natural world began in the tide pools on Cape Cod, and my desire to protect it started in the wetlands near my childhood home in Mt. Kisco, New York. Mom and Dad, thank you for taking me outdoors and for your endless support. You are, and will always be, my compass.

There are so many people—spanning decades and continents—who have made the main streets I have lived near or passed through feel like home, and I am indebted to them all.

To my high school teacher Heather Kranz, who nurtured my love for nature.

To George Bodarky and Julianne Welby for believing in me when I was just a freshman at Fordham University. You helped me find my voice and calling at WFUV.

To John Stossel, who gave me my first internship at ABC News.

To Winston Mitchell, who taught me what a "standup" was and helped me make a demo tape that sparked it all.

To Mary Young, who saw that demo tape and gave me my first job at KJCT News 8 in Grand Junction, Colorado. You launched what has since become an international journey of a lifetime.

To Bill Pohovey at WPLG in Miami and Susan Sullivan at WNBC in New York for letting me off the leash to explore.

To Stuart Emmrich and the editors of the *New York Times*' Travel section for giving me my first taste of international reporting early on.

To Tom Hundley and the team at the Pulitzer Center on Crisis Reporting for investing in climate stories before anyone else did.

To Chrystal Johns, Ingrid Ciprián-Matthews, Alison Pepper, Bill Felling, Kurt Davis, and David Rhodes—thank you for dispatching me to London for my first network job as a foreign correspondent with CBS News. And Katy Tur, thank you for giving me a guest room to sleep in (for much longer than I anticipated) while I found my footing.

To the London Bureau—Andy Clarke, Deb Thomson, Tina Kraus, Vicky Burston, Pete Gow, and Claire Day. Thank you for sending me to the farthest corners of the globe.

To Susan Zirinsky and Laurie Orlando for recognizing my passion for climate stories and reassigning me to a region saturated with them.

To the Los Angeles Bureau, including but not limited to Joelle Martinez, Mark Lima, Dell Alann, Kristen Weiser, Elli Fitzgerald, Bryan Keinz, and Sherri Sylvester—thank you for always entertaining, and helping research, my wild ideas.

To Adam Verdugo, Shawna Thomas, Rand Morrison, Mary Hager, Amy Rosner, Alturo Rhymes, Angel Canales, Kim Harvey, Brian Bingham, Peter Burgess, Chris Stover, Jon Tower, Norah O'Donnell, Gayle King, Tony Dokoupil, Nate Burleson, Jane Pauley, and Margaret Brennan for always making time for this journalism.

To Terri Stewart. You owe me so many hours of lost sleep, but I owe you so much more. Sincerely.

To Wendy McMahon and George Cheeks for leading a network that prioritizes this storytelling.

Many of the stories in *this* book began in the field alongside incredible producers. Thank you to Walker Dawson, Christian Duran, Barny Smith, Bill Applegate, Anam Siddiq, Chris Weicher, Kathleen Seccombe, Chris Spinder, Bob Kozberg, Simon Bouie, Jim Mietus, Somitra Butalia, Shanti Berg, Paul Facey, and Alyssa Estrada.

To the photojournalists who have helped bring each story in this book to life with humanity—Carlos Cortes, Luke Thomas, Michael Comfort, Roger Masterton, Gilbert Deiz, Scott Yun, Don Hale, Dave Lowther, Carlos Ascensio, John Weiser, Drew Morgan, and Larry Warner, to name just a few.

And to the army of engineers, fixers, security teams, and technical and editorial staff. None of this reporting would be possible without you.

To Rick Ramage for guiding my television career and to David Black for nurturing this written one. David, your belief in this book (and vision for it) made *me* believe in this book.

To One Signal and Julia Cheiffetz for your commitment to telling the untold and especially to my editor, Nick Ciani. Thank you for your effortless ability to refine and find the right words. Hannah Frankel, thank you for keeping me on deadline.

To the scientists, past and present, who have dedicated their lives to understanding our planet and helping others, including me, understand it too.

To the first responders on our climate front lines. You are all heroes.

To all the people I have met in the field, often during the worst times in their lives. Your vulnerability, strength, and grace under fire are the backbone of this book.

To my dear friends Leigh Kiniry, Kristin Fisher, and Rosemary Connors. Thank you for countless hours of scheming, collaborating, and shining a bright light on the truth. And to Joe Tringali, Sara Rosner, Pete Tierney, James Asensio, Jessie and Grant Nichols, Keith Paterson, Priscilla and Francisco Perez, Brian Kinsella, Eben and Suzan Hall, Patty Hamilton, Christian Zak, Bracken Feldman, Jennifer Georgiou, Jamie Yuccas, Alyssa Estrada, Kris Van Cleave, Janet Shamlian, Jen D'Cunha and Rayo Gomez, Alex and Brady Roper, and Ben Everin. Thank you for your listening ear over the years.

To Chocha and Uncle Eddie, who reared my imagination.

To Marc, Greg, Jessie, and Khristina for helping hold down the family fort in my absence.

To my nephews, Leo, Lucas, and Benno . . . and to my other "nephews," Jaime, Hugo, and Iggi. You will inherit this planet and help make it a better place for all to live. This I know to be true.

For Toast, who kept my feet warm as I wrote each morning at 5 a.m.

And to Ivan, my North Star, who brings out the light in others. Thank you for sharing and exploring this beautiful world with me.

Notes

Preface

xv Western Maui Community Wildfire Protection Plan: *Western Maui Community Wildfire Protection Plan*, Hawaii Wildfire Management Organization, 2014, https://dlnr.hawaii.gov/forestry/files/2023/08 /Western-Maui-CWPP14.pdf.

xvi June 12, 2014: "West Maui Community Wildfire Protection Plan Signed," *Lahaina News*, June 12, 2014, https://www.lahainanews.com /real-estate-features/2014/06/12/west-maui-community-wildfire-pro tection-plan-signed/.

Prologue

xxvi "liberated from the terrorists": "Syria's al-Assad Welcomes Fillon's Policies on Terrorism," Radio France Internationale (hereafter RFI), January 9, 2017, https://www.rfi.fr/en/france/20170109-syria-s-al -assad-welcomes-fillon-s-policies-terrorism.

xxvi 600,000 people had been killed: "Total Death Toll," Syrian Observatory for Human Rights, June 1, 2021, https://www.syriahr.com/en /217360/.

xxvii total crop failure: Colin P. Kelley, Shahrzad Mohtadi, Mark A. Cane, Richard Seager, and Yochanan Kushnir, "Climate Change in the Fertile Crescent and Implications of the Recent Syrian Drought," *Proceedings of the National Academy of Sciences* 112, no. 11 (2015): 3241–46. https:// doi.org/10.1073/pnas.1421533112.

xxvii "Down with the regime": Jamie Tarabay, "For Many Syrians, the Story of the War Began with Graffiti in Dara'a," CNN, March 15, 2018, https:// www.cnn.com/2018/03/15/middleeast/daraa-syria-seven-years-on-intl /index.html.

xxviii "armed conflict outbreaks": Lina Eklund, Ole Magnus Theisen, Matthias Baumann, Andreas Forø Tollefsen, Tobias Kuemmerle, and Jonas Østergaard Nielsen, "Societal Drought Vulnerability and the Syrian Climate-Conflict Nexus Are Better Explained by Agriculture than Meteorology," *Communications Earth & Environment* 3, no. 1 (April 2022), https://doi.org/10.1038/s43247-022-00405-w.

xxix United States was hammered by: Adam B. Smith, "2022 U.S. Billion-Dollar Weather and Climate Disasters in Historical Context," NOAA Climate.gov, January 10, 2023. https://www.ncei.noaa.gov/news/national-climate-202308.

xxix affected by environmental disasters: Hazard HQ Team, "2021 Climate Change Catastrophe Report," CoreLogic®, February 17, 2022, https://www.corelogic.com/intelligence/2021-climate-change-catastrophe-report/.

xxix reason for moving in 2022: Samantha Allen, "30% of Americans Cite Climate Change as a Motivator to Move in 2023," *Forbes*, July 4, 2023, https://www.forbes.com/home-improvement/features/americans-moving-climate-change/.

xxx 1,833 people were killed: Joan Brunkard, Gonza Namulanda, and Raoult Ratard, "Hurricane Katrina Deaths, Louisiana, 2005," *Disaster Medicine and Public Health Preparedness* 2, no. 4 (2008): 215–23, https://doi.org/10.1097/dmp.0b013e31818aaf55.

xxx more than one million people: Allison Plyer, "Facts for Features: Katrina Impact," The Data Center, August 26, 2016, https://www.datacenterresearch.org/data-resources/katrina/facts-for-impact/.

xxx "less supportive of spending to help the poor": Daniel J. Hopkins, "Flooded Communities: Explaining Local Reactions to the Post-Katrina Migrants," *Political Research Quarterly* 65, no. 2 (2012): 443–59, http://www.jstor.org/stable/41635245.

xxxi "outpacing the rights of white people": Robert Pape, *American Face of Insurrection: Analysis of Individuals Charged for Storming the US Capitol on January 6, 2021* (Chicago: Chicago Project on Security and Threats, 2022).

xxxi Jim Mulhern: Jim Mulhern, "NMPF Statement on Electoral-Vote Certification and Condemnation of Insurrection at U.S. Capitol," National Milk Producers Federation (NMPF), January 7, 2021, https://www.nmpf.org/nmpf-statement-on-electoral-vote-certification-and-condemnation-of-insurrection-at-u-s-capitol/.

The Flame Tamer

4 ignited by loggers and homesteaders: *The Great Fire of 1910*, United States Forest Service, n.d., https://www.fs.usda.gov/Internet/FSE_DOCUMENTS/stelprdb5444731.pdf.

5 4,000 soldiers to assist: Marvin Fletcher, "Army Fire Fighters," *Idaho Yesterdays* 16, no. 2 (Summer 1972), 12–15.

5 "Soot fell on the ice in Greenland": "The Great Fire of 1910," US Forest Service, https://www.fs.usda.gov/Internet/FSE_DOCUMENTS/stelprdb5444731.pdf.

Harry's War

7 freshly sparked flames: Charles E. Hardy, *The Gisborne Era of Forest Fire Research: Legacy of a Pioneer* (Washington, D.C.: United States Department of Agriculture, Forest Service, 1983).

Nature Fights Back

14 killed twenty-two people and destroyed more than 5,000 homes: CAL FIRE Sonoma-Lake-Napa Unit, "Tubbs Fire (Central LNU Complex)," Cal FIRE, October 25, 2019, https://www.fire.ca.gov/incidents/2017/10/8/tubbs-fire-central-lnu-complex/.

15 Scripps Institution of Oceanography: "FAQ: Climate Change in California," Scripps Institution of Oceanography, accessed September 15, 2023, https://scripps.ucsd.edu/research/climate-change-resources/faq-climate-change-california#:~:text=Average%20summer%20temperatures%20in%20California,occurring%20since%20the%20early%201970s.

15 twice as much land and hotter than they did in the early 2000s: James MacCarthy, Jessica Richter, Sasha Tyukavina, Mikaela Weisse, and Nancy Harris, "The Latest Data Confirms: Forest Fires Are Getting Worse," World Resources Institute, August 29, 2023, https://www.wri.org/insights/global-trends-forest-fires.

16 more than doubled: "Wildfire Climate Connection," National Oceanic and Atmospheric Administration (hereafter NOAA), accessed September 15, 2023, https://www.noaa.gov/noaa-wildfire/wildfire-climate-connection.

Nature's Hydrogen Bomb

22 "crafted no plan to evacuate": Paige St. John, Joseph Serna, and Rong-Gong Lin II, "Must Reads: Here's How Paradise Ignored Warnings

and Became a Deathtrap," *Los Angeles Times*, December 30, 2018, https://www.latimes.com/local/california/la-me-camp-fire-deathtrap-20181230-story.html.

Picking Up the Pieces

26 Ninety-five percent of the city was erased: "A Year Later, Debris Removal Completed for California's Deadliest Wildfire," CBS News, November 20, 2019, https://www.cbsnews.com/sacramento/news/a-year-later-debris-removal-completed-for-californias-deadliest-wildfire/.

Engineered Miracles

30 170 firefighters died battling western wildfires: *NWCG Report on Wildland Firefighter Fatalities in the United States: 2007–2016*, National Wildfire Coordinating Group (NWCG), December 2017, https://www.nwcg.gov/sites/default/files/publications/pms841.pdf.

"Go, Go, Go"

37 *Five percent*: Matthew C. Nisbet and Teresa Myers, "Trends: Twenty Years of Public Opinion about Global Warming," *The Public Opinion Quarterly* 71, no. 3 (2007): 444–70, http://www.jstor.org/stable/4500386.

37 $60 million a year: "Oil & Gas," OpenSecrets, accessed September 15, 2021, https://www.opensecrets.org/industries/indus.php?ind=e01.

37 University of Colorado's Media and Climate Change Observatory: M. Boykoff, L. Gifford, A. Nacu-Schmidt, and J. Osborne-Gowey, "US Television Coverage of Climate Change or Global Warming, 2004–2023, Media and Climate Change Observatory Data Sets," Cooperative Institute for Research in Environmental Sciences, University of Colorado, 2023, https://sciencepolicy.colorado.edu/icecaps/research/media_coverage/tv/index.html.

38 "issue of ethical concern": Laura W. Johnston and Frederick J. Ruf, "How 'An Inconvenient Truth' Expanded the Climate Change Dialogue and Reignited an Ethical Purpose in the United States," Digital Georgetown, Georgetown University, 2013, https://repository.library.georgetown.edu/handle/10822/558371.

38 wind and solar have spiked 90 percent: "Renewable Energy," Center for Climate and Energy Solutions, August 28, 2023, https://www.c2es.org/content/renewable-energy/.

40 33 people were killed and 11,116 homes, businesses, and other structures

were destroyed: "2020 Incident Archive," CAL FIRE, accessed September 15, 2023, https://www.fire.ca.gov/incidents/2020.

The Whack-A-Mole Defensive

42 12 percent of the state: Clarke A. Knight, Lysanna Anderson, M. Jane Bunting, Marie Champagne, Rosie M. Clayburn, Jeffrey N. Crawford, Anna Klimaszewski-Patterson, et al., "Land Management Explains Major Trends in Forest Structure and Composition over the Last Millennium in California's Klamath Mountains," *Proceedings of the National Academy of Sciences* 119, no. 12 (2022), https://doi.org/10.1073/pnas.2116264119.

42 "Wawona": "What's in a Name?," National Parks Service, September 18, 2014, https://www.nps.gov/yose/blogs/whats-in-a-name.htm.

43 "act with vindictive earnestness": Gilbert King, "Where the Buffalo No Longer Roamed," *Smithsonian*, July 17, 2012, https://www.smithsonianmag.com/history/where-the-buffalo-no-longer-roamed-3067904/.

43 100,000 were displaced: Jessica Wolf, "Revealing the History of Genocide against California's Native Americans," University of California, Los Angeles (UCLA), June 1, 2022, https://newsroom.ucla.edu/stories/revealing-the-history-of-genocide-against-californias-native-americans.

45 "60 years is about 100 times faster": Rebecca Lindsey, "Climate Change: Atmospheric Carbon Dioxide," NOAA Climate.gov, May 12, 2023, https://www.climate.gov/news-features/understanding-climate/climate-change-atmospheric-carbon-dioxide.

When Giants Fall

48 Castle Fire: "Wildfires Kill Unprecedented Numbers of Large Sequoia Trees," National Parks Service, July 18, 2023, https://www.nps.gov/articles/000/wildfires-kill-unprecedented-numbers-of-large-sequoia-trees.htm.

To Stay or to Go

58 First Street Foundation: Alex Wigglesworth, "California Properties at Risk of Wildfire Expected to See Sixfold Increase in 30 Years," *Los Angeles Times*, May 16, 2023, https://www.latimes.com/california/story/2022-05-16/california-properties-at-risk-of-wildfire-expected-to-grow.

59 forests, grasslands, and shrublands: Volker C. Radeloff, David P. Helmers, Miranda H. Mockrin, Amanda R. Carlson, Todd J. Hawbaker, and

Sebastián Martinuzzi, "The 1990–2020 Wildland-Urban Interface of the Conterminous United States—Geospatial Data," Forest Service Research Data Archive, 2022, https://doi.org/10.2737/rds-2015-0012-3.

"Nature's Way"

61 13 percent survival rate in the Camp Fire: Paige St. John, Joseph Serna, and Rong-Gong Lin II, "Must Reads: Here's How Paradise Ignored Warnings and Became a Deathtrap," *Los Angeles Times,* December 30, 2018, https://www.latimes.com/local/california/la-me-camp-fire-death trap-20181230-story.html.

61 according to a 2019 report: "UN Report: Nature's Dangerous Decline 'Unprecedented'; Species Extinction Rates 'Accelerating'—United Nations Sustainable Development," United Nations, May 6, 2019, https://www.un.org/sustainabledevelopment/blog/2019/05/nature -decline-unprecedented-report/.

62 U.S. Fish and Wildlife Service: Julia Pinnix, ed., "Beavers Work to Improve Habitat," U.S. Fish & Wildlife Service, September 14, 2023, https://www.fws.gov/story/beavers-work-improve-habitat.

64 vanishing from the landscape: California Wetland Monitoring, "How Much Wetland Area Has California Lost?," My Water Quality: Are Our Wetlands Healthy? April 20, 2016, https://mywaterquality.ca.gov /eco_health/wetlands/extent/loss.html.

65 allocates $300 million: "Measure W," Los Angeles Area Chamber of Commerce, accessed September 15, 2023, https://lachamber.com /advocacy-endorsements/measure-w/.

65 land for conservation efforts: Hayley Smith, "L.A. Promised to Stop Wasting so Much Stormwater. But Progress Has Been Painfully Slow," *Los Angeles Times,* February 21, 2023, https://www.latimes.com/california /story/2023-02-21/progress-on-l-a-stormwater-capture-program-is-slowing.

Ice Chaser

75 "along most coasts": J. H. Mercer, "West Antarctic Ice Sheet and CO_2 Greenhouse Effect: A Threat of Disaster," *Nature* 271, no. 5643 (1978): 321–25. https://doi.org/10.1038/271321a0.

75 "fossil fuels to other sources of energy": Mercer, 321–25.

76 *New Scientist:* James Hansen, "Huge Sea Level Rises Are Coming—Unless We Act Now," *New Scientist,* July 25, 2007, https://www.newscientist.com/article /mg19526141-600-huge-sea-level-rises-are-coming-unless-we-act-now/.

Life Before

85 80 to 85 percent of all major Category 3, 4, and 5 hurricanes: "First Cabo Verde Missions Explore Earliest Roots of Atlantic Hurricanes," Communication Services, NOAA's Atlantic Laboratory, September 14, 2022, https://www.aoml.noaa.gov/first-cabo-verde-missions/.

Answers on Ice

89 279 billion metric tons: "Greenland Ice Mass Loss 2002–2023—Climate Change," NASA Global Climate Change: Vital Signs of the Planet, August 23, 2023, https://climate.nasa.gov/climate_resources/264/video -greenland-ice-mass-loss-2002-2023/.

94 Our oceans have absorbed 90 percent: Rebecca Lindsey and LuAnn Dahlman, "Climate Change: Ocean Heat Content," NOAA Climate .gov, September 6, 2023, https://www.climate.gov/news-features /understanding-climate/climate-change-ocean-heat-content.

94 conveyor belt by 2050: Peter Ditlevsen and Susanne Ditlevsen. "Warning of a Forthcoming Collapse of the Atlantic Meridional Overturning Circulation," *Nature Communications* 14, no. 1 (2023), https://doi .org/10.1038/s41467-023-39810-w.

94 8 percent a decade: James P. Kossin, Kenneth R. Knapp, Timothy L. Olander, and Christopher S. Velden, "Global Increase in Major Tropical Cyclone Exceedance Probability over the Past Four Decades," *Proceedings of the National Academy of Sciences* 117, no. 22 (2020): 11975–80, https://doi.org/10.1073/pnas.1920849117.

97 70 percent of military housing: "Corvias Repairs 2,600+ Homes at Fort Polk after Devastating Storms," Corvias, May 21, 2021, https:// www.corvias.com/news/corvias-repairs-2600-homes-fort-polk-after -devastating-storms.

Regulating Sandcastles

110 27,000 square miles: "1927 Flood Photograph Collection," MS Digital Archives, Mississippi Department of Archives and History, accessed September 15, 2023, https://da.mdah.ms.gov/series/1927flood.

112 "predict future conditions": John Muyskens and Samuel Oakford, "Extreme Floods Expose the Flaws in FEMA's Risk Maps," *Washington Post*, December 6, 2022, https://www.washingtonpost.com/climate-en vironment/interactive/2022/fema-flood-risk-maps-failures/.

112 "NFIP flood insurance claims": "Facts about the National Flood Insurance

Program (NFIP) Flood Insurance," FEMA, September 14, 2020, https://www.fema.gov/blog/facts-about-national-flood-insurance-program-nfip-flood-insurance.

112 flood risk variables: "NFIP's Pricing Approach," FEMA, accessed September 15, 2023, https://www.fema.gov/flood-insurance/risk-rating.

113 mortgages are federally backed: "Covid-19 Housing Protections: Mortgage Forbearance and Other Federal Efforts Have Reduced Default and Foreclosure Risks," U.S. Government Accountability Office, July 20, 2021, https://www.gao.gov/products/gao-21-554.

114 26 percent more often: "Pioneering Study Shows Flood Risks Can Still Be Considerably Reduced If All Global Promises to Cut Carbon Emissions Are Kept," University of Bristol, March 7, 2023, https://www.bristol.ac.uk/news/2023/march/flood-risk.html.

Lessons Learned

120 all occurred in the past twenty years: "Costliest U.S. Tropical Cyclones," NOAA's National Centers for Environmental Information, accessed September 11, 2023, https://www.ncei.noaa.gov/access/billions/dcmi.pdf?itid=lk_inline_enhanced-template.

120 26 people per square mile: Aon Benfield, *Weather, Climate & Catastrophe Insight, 2017 Annual Report*, Aon, 2018, https://www.aon.com/spain/temas-destacados/Annual-report-weather-climate-2017.pdf.

120 the first and only study of its kind: Oliver E. Wing, Paul D. Bates, Andrew M. Smith, Christopher C. Sampson, Kris A. Johnson, Joseph Fargione, and Philip Morefield, "Estimates of Present and Future Flood Risk in the Conterminous United States," *Environmental Research Letters* 13, no. 3 (2018): 034023, https://doi.org/10.1088/1748-9326/aaac65.

121 75 percent of the funds needed: "Understanding Individual Assistance and Public Assistance," FEMA, October 12, 2017, https://www.fema.gov/press-release/20230425/understanding-individual-assistance-and-public-assistance.

122 sea level rise by 2045: *Monroe County Roadway Vulnerability Study Final Report*, Monroe County, Florida, August 2022, https://monroecounty-fl.gov/DocumentCenter/View/31905/MCRVS-Final-Report?bidId=.

Asleep at the Wheel

126 taxpayer dollars to help buy homes: "Bipartisan Infrastructure Law Provides Historic Levels of Funding for Resilience Projects Nationwide,"

FEMA, May 19, 2023, https://www.fema.gov/fact-sheet/bipartisan-infrastructure-law-provides-historic-levels-funding-resilience-projects.

Foote's Notes

134 when our oceans were up to 82 feet higher: Rebecca Lindsey, "Climate Change: Atmospheric Carbon Dioxide," NOAA Climate.gov, May 12, 2023, https://www.climate.gov/news-features/understanding-climate/climate-change-atmospheric-carbon-dioxide.

HWY 89

161 "Land of the Burning Ground": Richard Grant, "The Lost History of Yellowstone," *Smithsonian*, January 2021, https://www.smithsonianmag.com/history/lost-history-yellowstone-180976518/.

161 United States Geological Survey: Yellowstone Volcano Observatory, "Colter's Hell: Tales of the First European-American to Step Foot in Yellowstone," U.S. Geological Survey, May 20, 2019, https://www.usgs.gov/observatories/yvo/news/colters-hell-tales-first-european-american-step-foot-yellowstone.

164 Global Carbon Project: Pierre Friedlingstein, Michael O'Sullivan, Matthew W. Jones, Robbie M. Andrew, Luke Gregor, Judith Hauck, Corinne Le Quéré, et al. "Global Carbon Budget 2022," *Earth System Science Data* 14, no. 11 (2022): 4811–4900, https://doi.org/10.5194/essd-14-4811-2022.

164 total CO_2 output: "Sources of Greenhouse Gas Emissions," EPA, August 25, 2023, https://www.epa.gov/ghgemissions/sources-greenhouse-gas-emissions.

172 on where tornadoes formed: Vittorio A. Gensini and Harold E. Brooks, "Spatial Trends in United States Tornado Frequency," *Nature* Partner Journals *Climate and Atmospheric Science* 1, no. 1 (2018), https://doi.org/10.1038/s41612-018-0048-2.

173 one step further: Walker S. Ashley, Alex M. Haberlie, and Vittorio A. Gensini, "The Future of Supercells in the United States," *Bulletin of the American Meteorological Society* 104, no. 1 (2023), https://doi.org/10.1175/bams-d-22-0027.1.

The Land Between the Rivers

176 "must be our Mississippi": Claudio Saunt, *Unworthy Republic: The Dispossession of Native Americans and the Road to Indian Territory* (New York: W. W. Norton & Company, 2020).

All Hands on Deck

185 $25,000 of damage: "Myths vs Facts: The True Cost of Flooding," FEMA, January 1, 2022, https://community.fema.gov/PreparednessConnect/s /article/Myths-vs-Facts-The-True-Cost-of-Flooding.

Ms. Becky's Place

187 15 people were killed: Matt Hughes, "City, County Looking at Future of Industry in Dawson Springs," *Times Leader*, July 2, 2022, https://www .timesleader.net/news/city-county-looking-at-future-of-industry-in -dawson-springs/article_189359fc-b908-5371-abba-5b006fd91219.html.

Dexter Dirt

204 2011 and 2015: Patrick Keaton, *Selected Statistics from the Common Core of Data: School Year 2011–12: First Look*, National Center for Education Statistics, October 2023, https://files.eric.ed.gov/fulltext /ED544222.pdf.

204 the same time period: Alana Semuels, "'They're Trying to Wipe Us Off the Map,' Small American Farmers Are Nearing Extinction," *Time*, November 27, 2019, https://time.com/5736789/small-american-farm ers-debt-crisis-extinction/.

204 populations decline since 2010: Clark Kauffman, "Iowa Remains a Less Diverse State, as Two-Thirds of Its Counties Lose Population," *Iowa Capital Dispatch*, August 12, 2021, https://iowacapitaldispatch .com/2021/08/12/iowa-remains-a-less-diverse-state-as-two-thirds-of -its-counties-lose-population/.

204 200,000 small farms have shuttered: "Farming and Farm Income," Economic Research Service, U.S. Department of Agriculture, August 31, 2023, https://www.ers.usda.gov/data-products/ag-and-food-statistics -charting-the-essentials/farming-and-farm-income/.

205 where most of us live: "Nation's Urban and Rural Populations Shift Following 2020 Census," U.S. Census Bureau, December 9, 2022, https://www.census.gov/newsroom/press-releases/2022/urban -rural-populations.html.

205 to only $3 billion in 2018: Chad P. Bown, "China Bought None of the Extra $200 Billion of US Exports in Trump's Trade Deal," Peterson Institute for International Economics, July 19, 2022, https://www.piie .com/blogs/realtime-economics/china-bought-none-extra-200-billion-us -exports-trumps-trade-deal.

206 crop insurance payments: Georgina Gustin, "US Taxpayers Are Spending Billions on Crop Insurance Premiums to Prop up Farmers on Frequently Flooded, Unproductive Land," *Inside Climate News*, March 30, 2022, https://insideclimatenews.org/news/30032022/crop-insurance-premiums-flood-agriculture/.

206 Union of Concerned Scientists: "The Hidden Costs of Industrial Agriculture," Union of Concerned Scientists, August 24, 2008, https://www.ucsusa.org/resources/hidden-costs-industrial-agriculture.

207 July 2018 to 2019: John Newton, "Farm Bankruptcies Rise Again," American Farm Bureau Federation, October 30, 2019, https://www.fb.org/market-intel/farm-bankruptcies-rise-again.

207 20 percent of the country's food: Christine Whitt, "A Look at America's Family Farms," USDA, January 23, 2020, https://www.usda.gov/media/blog/2020/01/23/look-americas-family-farms.

Creampuff Pioneers

209 3.5 million eco-refugees: Jayne Karsten, "Part IV: The Dust Bowl Migrants," The Kennedy Center, November 4, 2019, https://www.kennedy-center.org/education/education-sandbox/education/resources-for-educators/classroom-resources/media-and-interactives/media/literary-arts/john-steinbeck--the-grapes-of-wrath/chapters/the-dust-bowl-migrants/.

212 "EDUCATIONAL FACILITIES FOR SEASON DOUBTFUL": Report, 79 Congressional Record: Proceedings and Debates of the United States Congress. Volume 79, Part 9 (1935).

212 "creampuff pioneers": Congressional Record: Proceedings and Debates of the United States Congress, vol. 111, part 29 (Washington, D.C.: U.S. Government Printing Office, 1965).

217 4 degrees since 1970: D. R. Reidmiller, C. W. Avery, D. R. Easterling, K. E. Kunkel, K. L. M. Lewis, T. K. Maycock, and B. C. Stewart, "Impacts, Risks, and Adaptations in the United States: Fourth National Climate Assessment, Volume II," NOAA U.S. National Climate Assessment, 2017, doi: 10.7930/NCA4.2018

Checkmate

221 took their lives between 2014 and 2018: Katie Wedell, Lucille Sherman, and Sky Chadde, "Midwest Farmers Face a Crisis. Hundreds Are Dying by Suicide," *USA Today*, March 9, 2020, https://www.usatoday.com

/in-depth/news/investigations/2020/03/09/climate-tariffs-debt-and
-isolation-drive-some-farmers-suicide/4955865002/.

Turning the (Dinner) Tables

234 driest in 128 years: Haidong Zhao, Lina Zhang, M. B. Kirkham, Stephen
M. Welch, John W. Nielsen-Gammon, Guihua Bai, Jiebo Luo, et al.,
"U.S. Winter Wheat Yield Loss Attributed to Compound Hot-Dry-
Windy Events," *Nature Communications* 13, no. 1 (2022), https://doi
.org/10.1038/s41467-022-34947-6.

234 even drier than during the Dust Bowl: Mitch Smith, "First Scorched,
Then Soaked: Weather Whiplash Confounds Farmers," *New York Times*,
August 9, 2023, https://www.nytimes.com/2023/08/09/us/kansas
-wheat-harvest-drought.html.

234 70 percent of their harvest: Suman Naishadham and Eric Gay, "Drought
Takes Toll on Country's Largest Cotton Producer," AP News, October 7,
2022, https://apnews.com/article/business-texas-droughts-agricul
ture-us-department-of-42112182a5fd6a1b5cf8e71aa5897326#.

234 rice harvest shrunk by 50 percent: Michelle Bandur, "2022 Was a
Bad Year for Rice Harvesting in California, Producing Only Half the
Usual Amount." KCRA 3, September 14, 2022, https://www.kcra.com
/article/2022-california-rice-harvest-drought-supply/41200588#.

234 decline up to 30 percent by 2050: *Adapt Now: A Global Call for
Leadership on Climate Resilience*, Global Commission on Adaptation,
September 13, 2019, https://gca.org/wp-content/uploads/2019/09
/GlobalCommission_Report_FINAL.pdf.

Index

About the Author

Jonathan Vigliotti is an Emmy and Edward R. Murrow Award–winning CBS News national correspondent whose work has appeared on numerous platforms including CBS *Sunday Morning, Face the Nation,* the CBS *Evening News,* and more. His reporting has taken him to more than forty countries and territories across six continents. Follow him on Instagram @JonathanVigliotti. *Before It's Gone* is his first book.